Sensorama

Sensorama

A phenomenalist analysis of spacetime and its contents

Michael Pelczar

OXFORD
UNIVERSITY PRESS

OXFORD
UNIVERSITY PRESS

Great Clarendon Street, Oxford, OX2 6DP,
United Kingdom

Oxford University Press is a department of the University of Oxford.
It furthers the University's objective of excellence in research, scholarship,
and education by publishing worldwide. Oxford is a registered trade mark of
Oxford University Press in the UK and in certain other countries

First Edition published in 2015

Published in the United States of America by Oxford University Press
198 Madison Avenue, New York, NY 10016, United States of America

British Library Cataloguing in Publication Data
Data available

Library of Congress Control Number: 2014947816

ISBN 978-0-19-873265-5

Slight not what's near through aiming at what's far.

Euripides

Preface

The central thesis of this book is that the best way to fit together what we know about time, space, and consciousness is through a phenomenalist metaphysics that reduces facts about spacetime and its contents to phenomenological facts—facts about the phenomenal character of actual and potential experience. On this view, consciousness itself exists in neither time nor space, but serves as a basis for the logical construction of all spatiotemporal phenomena, including physical phenomena; conscious properties (qualia) are proto-spatiotemporal properties whose instances are broadly constitutive of, rather than contained by, spacetime.

I took the first step toward this view six years ago, while thinking about what would happen, in the limit, as you sped up the neural processing associated with some kind of conscious experience. My guess was that nothing would happen at all, phenomenologically speaking. This brought into view the possibility that a durationless event could have the subjective character of an ordinary experience of duration or change. Further reflection led me to conclude, against my expectations, that this might be more than a mere logical possibility, but a reality that applies to our own mental lives.

I am a long-time admirer of Leibniz. I especially admire the fact that Leibniz, alone in the history of modern philosophy, had the audacity to suggest an idealist reduction not just of space and the spatial, but of time and the temporal as well.

Regrettably, Leibniz does not clearly explain how the reduction of time is supposed to go. In particular, he does not address the objection that an idealist reduction of time is impossible, since consciousness is itself extended in time. But this objection loses its force, if, as I had concluded, it is coherent to suppose that our own experiences of temporal succession and change lack temporal extent.

Having surmised that a conscious experience could present a subjective appearance of change without having any objective duration, it seemed reasonable to ask whether we have any phenomenological basis for thinking that conscious experience exists in time at all. Certainly, it could no longer be argued that experiences are shown to exist in time by the fact that many of them present subjective appearances of change. If there was a phenomenological reason to keep consciousness in time, it must have to do not with individual conscious experiences, but with the way in which individual experiences join up to form whole streams of consciousness.

Here again my thinking took an unexpected turn. I began by asking whether it was possible to square the existence of streams of consciousness with an atemporal conception of experience; I ended by concluding that this is not only possible, but even necessary, if we are to avoid committing ourselves to a rather extraordinary dualism of mind and body.

These thoughts appeared not only to remove the greatest conceptual barriers to a theory like Leibniz's, but also to provide a clear motive for pursuing such a theory. Given that we have no good phenomenological reason to think that consciousness exists in spacetime, and some incentive to think that it does not, the question naturally arises: could it be that conscious experience does not, in fact, exist in time or space? And this immediately raises the further question: if conscious experiences do not exist in spacetime, then how do they relate to the things that do? Any answer must take the form of a theory that sees the relationship between consciousness and the physical world as other than temporal, and among the theories that this leaves us to consider, a thoroughgoing phenomenalism is by far the most attractive—or so I am going to argue.

In working out the ideas for this book, I have had the good fortune to discuss them at length with the philosophers of Singapore. These discussions occurred mainly in the course two reading groups. The first of these took place just as I was getting started on the manuscript, and included Ben Blumson, Neil Sinhababu, Kyle Swan, and Weng Hong Tang; it would not be too much to say that the feedback I received from that group made this book possible. The second reading group took place toward the end of the writing process, and included Mark D'Cruz, Axel Gelfert, Neil Mehta, and Nico Silins, in addition to most of the participants of the earlier group. If the first reading group made the book possible, the second made it readable. I am grateful to the participants of both groups for their sharp and good-natured criticism, which has improved the book in almost every way.

I have also had the benefit of fruitful conversations with various visitors to Singapore, including Roger Ames, John Baez, Tim Bayne, David Bourget, Arindam Chakrabarti, David Chalmers, Philippe Chuard, John Greenwood, Frank Jackson, Uriah Kriegel, Franklin Perkins, Huw Price, and Brook Ziporyn. A sporadic but stimulating correspondence with Barry Dainton provided much-needed encouragement at several crucial junctures. Walter Ott's comments on an early draft corrected some important historical errors; Jay Garfield and Hui-chieh Loy saved me from errors of another sort. John Williams and Eric Thompson served as invaluable sounding boards from the earliest stages of the project. Various ideas related to the book were the subject of presentations to philosophical audiences at the Chinese University of Hong Kong, the University of Wyoming, the Massachusetts Institute of Technology, the University of Virginia, and the

University of Mississippi. I received helpful criticism on each occasion, particularly from Jim Cargile, Brie Gertler, Caspar Hare, Harold Langsam, Neil Manson, Trenton Merricks, Marc Moffett, and Steven Skultety.

Last, but certainly not least, the book has benefited greatly from insightful and constructive comments from the anonymous readers who reviewed the manuscript for Oxford University Press.

Michael W. Pelczar

Singapore,
June 2014

Contents

1. Introduction 1
   1.1 The Big Picture 1
   1.2 A forgotten chapter of metaphysics 3
   1.3 Terminology 7
   1.4 Introspection 10
   1.5 A glance ahead 27

2. Space, time, and spacetime 33
   2.1 Introduction 33
   2.2 Distance, duration, and simultaneity 36
   2.3 Inertial coordinate systems 40
   2.4 Relativistic spacetime 44
   2.5 Conclusion 54

3. Phenomenal duration, succession, and change 56
   3.1 Introduction 56
   3.2 The well-foundedness of experience 59
   3.3 The ubiquity of phenomenal succession 62
   3.4 The general case 66
   3.5 Conclusion 68

4. Phenomenal simultaneity 69
   4.1 Introduction 69
   4.2 Russell's Thesis 70
   4.3 The unity of the Now 73
   4.4 Russellian dualism 79
   4.5 Conclusion 82

5. The stream of consciousness 84
   5.1 Introduction 84
   5.2 The paradox of the stream of consciousness 87
   5.3 A temporal solution 92
   5.4 An atemporal solution 101
   5.5 Phenomenal sorites 106
   5.6 Conclusion 110

6. Idealism 112
   6.1 Introduction 112
   6.2 Kant's metaphysics of the thing-in-itself 120
   6.3 Leibniz's monadology 122

6.4 Mill's permanent possibilities of sensation 124
6.5 Conclusion 126

7. Phenomenalism: A first appraisal 127
7.1 Introduction 127
7.2 Mind-dependence 128
7.3 Ultimate mysteries 130
7.4 The veil of ideas 136
7.5 Conclusion 137

8. Radical phenomenalism 139
8.1 Introduction 139
8.2 Lessons from Mill 140
8.3 The central theory 146
8.4 Solving the extra-bodies problem 155
8.5 Conclusion 159

9. Idealism vindicated 160
9.1 Introduction 160
9.2 Unusual origins 160
9.3 Deceptive appearances 165
9.4 The imperceptible 167
9.5 Conclusion 176

10. The conscious self 178
10.1 Introduction 178
10.2 No self? 180
10.3 Reductive theories of the conscious self 181
10.4 Particularism about the self 189
10.5 Conclusion 192

11. Intentionality 194
11.1 Introduction 194
11.2 Phenomenalistic intentionality 196
11.3 Internalism versus externalism 198
11.4 Conclusion 206

12. Conclusion 207

Appendix: objections to relativity 215
Bibliography 225
Index 235

1

Introduction

> Once Einstein said that the problem of the Now worried him seriously. He explained that the experience of the Now means something special for man, something essentially different from the past and the future, but that this important difference does not and cannot occur within physics. That this experience cannot be grasped by science seemed to him a matter of painful but inevitable resignation. I remarked that all that occurs objectively can be described in science; on the one hand the temporal sequence of events is described in physics; and, on the other hand, the peculiarities of man's experiences with respect to time, including his different attitude towards past, present, and future, can be described and (in principle) explained in psychology. But Einstein thought that these scientific descriptions cannot possibly satisfy our human needs; that there is something essential about the Now which is just outside the realm of science.
>
> Rudolf Carnap, *Intellectual Autobiography*

1.1 The Big Picture

There was a time, not so very long ago, when the mind–body debate was primarily a conversation between dualists and idealists. Then things changed. I want them to change back.

Today, a majority of professional philosophers believe that some kind of physicalism holds the best prospects for a satisfying solution to the mind–body problem. I disagree; I think that the best prospects for a satisfying solution lie with a broadly idealist metaphysics. This is not because I think that physicalism is a dehumanizing world-view (it isn't), nor because I think that idealism brings us nearer to God (it doesn't). It is because I think that a certain kind of idealism—more specifically, a certain kind of *phenomenalism*—does the best job of fitting together what we know about space, time, and consciousness.

According to the theory I favor, facts about spacetime and its contents—facts that verify various descriptions of spatial or temporal states of affairs, or various assertions about the fundamental structure of spacetime—reduce to broadly phenomenological facts about conscious experience.

I call the theory *radical phenomenalism*. The theory is a kind of phenomenalism, because it takes facts about conscious experience as metaphysically fundamental. It deserves to be called radical because unlike traditional phenomenalist theories, it aspires to understand not just space and the spatial, but also time and the temporal, in purely phenomenological terms. This is in contrast to the phenomenalism of J.S. Mill, which takes conscious experience to be essentially temporal, and therefore unsuitable as a basis for the wholesale reduction of temporal states of affairs.

The theory invites a number of obvious objections:

(1) We have direct introspective evidence that our experiences persist and change over time; radical phenomenalism is inconsistent with this.
(2) We have direct introspective evidence that various experiences have various qualities simultaneously; this also conflicts with radical phenomenalism.
(3) Radical phenomenalism is inconsistent with the fact that many of our experiences occur as parts of temporally extended streams of consciousness.
(4) Radical phenomenalism cannot give a plausible account of the relationship between conscious experiences and physical events (some would go as far as to say that a radical phenomenalist cannot even give an *intelligible* account of this relationship).

I address these objections in Chapters 3, 4, 5, and 6–9, respectively. To anticipate: in Chapter 3, I argue that introspection gives us no basis to suppose that our conscious experiences are characterized by any duration or succession in physical time (or spacetime). In Chapter 4, I argue that introspection gives us no reason to think that our experiences are characterized by simultaneity in physical time (and so, for example, no reason to think that any of our experiences is simultaneously an experience as of something round and an experience as of something red). In Chapter 5, I grant that experiences typically occur as parts of streams of consciousness, but deny that there is anything in the intrinsic nature of these streams to compel us to think of them as extended in time.

That covers the first three objections. The challenge of overcoming objection (4) is tantamount to that of defending a broadly idealistic account of the relationship between consciousness and the physical world. Only such an account can make the relationship intelligible, on the assumption that the relationship is neither spatial nor temporal. This large task is the business of the central chapters of the book, wherein I characterize idealism in general terms and identify phenomenalism as a species of idealism (Chapter 6), argue that phenomenalist versions of idealism are superior to non-phenomenalist versions (Chapter 7), develop a modern version of phenomenalism, distinguished by its assumption of a timeless domain of conscious experience (Chapter 8), and address the main objections to idealism in general (Chapter 9).

1.2 A forgotten chapter of metaphysics

As serious as the objections enumerated above are, there is an even greater obstacle to persuading anyone of the attractiveness, or even the consideration-worthiness, of radical phenomenalism.

David Lewis once reported that the most common reaction to his avowal of modal realism—the view that every possible world exists *in concreto*—was not any well-thought-out objection, but rather an "incredulous stare":

> What arguments can be given against realism about possible worlds? I have met with few arguments—incredulous stares are more common.[1]

Lewis went on to address the few objections to his position that he had encountered, but declined to engage with the incredulous stares. It is tempting to follow his example in the present case, focusing on well-formulated objections to radical phenomenalism, rather than on the incredulous stares that advocacy of this theory elicits.

This is a temptation that I am going to resist. The purpose of this book is not merely to record my own thoughts, but to start a conversation, and to do that, I must find some way past the incredulous stare.

Here I am in a better position than Lewis. For I am not the first to think that there are problems with the common view that consciousness exists in the same time as physical events, and not the first to suggest that the best solution to these problems is to construe conscious experience as the logical foundation of spacetime and its contents. Around the beginning of the last century, developments in the scientific understanding of time and space prompted Henri Poincaré and Bertrand Russell to re-examine the relationship between consciousness and the physical world.

In 1896, Poincaré prepared the ground for a forthcoming scientific revolution, with an essay now widely regarded as containing the seeds of the relativistic conception of time subsequently brought to fruition by Einstein. Less well known is the fact that Poincaré saw that such a conception of time appeared to have profound consequences for the philosophy of mind. On one hand,

[1] Lewis (1973), 86.

When we say that two conscious facts are simultaneous we mean that they profoundly interpenetrate, so that analysis cannot separate them without mutilating them. The order in which we arrange conscious phenomena does not admit of any arbitrariness. It is imposed upon us and of it we can change nothing.[2]

On the other hand,

We have no direct intuition of simultaneity, nor of the equality of two durations. If we think we have such an intuition, we are deluded. What we have are certain rules which we apply almost always without even thinking about them . . . These rules are not imposed upon us and we might amuse ourselves by inventing others; but they could not be cast aside without greatly complicating the enunciation of the laws of physics, mechanics and astronomy . . . We therefore choose these rules, not because they are true, but because they are the most convenient.[3]

In other words, even though the succession or simultaneity of two physical events is a matter of convention, contingent on a pragmatic choice of rules in terms of which to describe those events, there is nothing conventional or pragmatic about the succession or simultaneity of two events occurring as parts of one's own conscious mental life.

From this, Poincaré concludes that conscious experience does not exist in physical time at all. Rather, when we apply temporal concepts to our own experience, we use those concepts to pick out features of experience that are distinct from, and logically prior to, persistence, succession, and change in physical time (or spacetime):

In a word, psychologic time is given to us and must create scientific and physical time.[4]

In short, Poincaré proposes to reconcile the apparent conflict between physical and "psychologic" time with a sweeping idealism that reduces facts about spacetime and its contents to facts about conscious experience.

Poincaré's startling proposal has gone almost completely unremarked in the hundred-odd years since he made it.[5]

[2] Poincaré (1913), 223. [3] Poincaré (1913), 234. [4] Poincaré (1913), 227.

[5] A notable exception is Lee (2007), which is to my knowledge the only recent attempt to assess the implications of a relativistic conception of spacetime for the study of consciousness. The general thrust of Lee's discussion is that the relativistic structure of spacetime forces us to recognize that the relationship between the phenomenology of experience and the objective temporal structure of experience is much looser than we are prone to think. It seems to me that this basic point is correct, and I shall offer my own support for it in Chapter 3. However, Lee and I take different lessons from the basic point. Lee concludes that although consciousness exists in physical time, the phenomenal character of an experience does not depend on facts about the objective temporal ordering of the qualia that it instantiates. I conclude that *if* consciousness exists in physical time, then the phenomenal characters of some, but not all, experiences depend on facts about the objective temporal ordering of qualia-instantiations. Of course, my considered view is that qualia do not get

However, Poincaré is not the only one to have noticed that relativity seems to have implications for the philosophy of mind. In a neglected essay from 1914, Bertrand Russell contends that a subjective conscious appearance of simultaneous states of affairs, such as a visual perception of a lime being simultaneously green and roundish, is an absolutely simultaneous instantiation of qualia (phenomenal greenness, phenomenal roundishness), and therefore, given relativity, a singular point-event in spacetime:

> Two events which are simultaneous in my experience may be spatially separate in psychical space, e.g. when I see two stars at once. But in physical space these two events are not separated, and indeed they occur in the same place in space-time. Thus in this respect relativity theory has complicated the relation between perception and physics.[6]

While relativity (or something like it) leads Poincaré to deny consciousness location in spacetime, it leads Russell to the view that conscious experiences, or at least our most basic conscious experiences, have spatiotemporal location but no spatiotemporal extent (or no more extent than what a single point of spacetime possesses). According to Russell, the elemental manifestations of consciousness are singular point-events.

It will be useful to have labels for the positions that Poincaré and Russell take on the status of consciousness in relation to physical time. Let us call Poincaré's view that conscious experience does not exist in time *phenomenological atemporalism* or "atemporalism," for short, and call the opposing view, that is, the standard view that conscious experiences do occur in physical spacetime (or at least in physical time) *phenomenological temporalism*. And let us use the term "pointillism" to refer to Russell's view that individual conscious experiences (or at least the simplest experiences) are spatiotemporally located but spatiotemporally unextended events.

What led Poincaré and Russell to these extraordinary positions? How, exactly, is the modern scientific understanding of time and space supposed to "complicate the relation between perception and physics"?

To begin with, there is the fact, or alleged fact, that conscious experiences have determinate durations. When I have a throbbing headache, each throb of pain lasts for a certain amount of time, as do the intervals between throbs. The duration of a throb is not something that we are at liberty to describe as longer or shorter, according to our preferences. There is an absolute, objective fact of the

instantiated in time at all; but if they did, I see no reason why the objective temporal order of their instantiation couldn't play a role in the phenomenal character of a conscious experience. For a view similar to Lee's, see Zuboff (1977–8) and Zuboff (1981).

[6] Russell (1914), 130.

matter about the duration of each throb, which is what makes this headache more unpleasant than some, and less unpleasant than others. Likewise, when I listen to music, each note occurs in my consciousness a certain amount of time before or after the notes that immediately precede or follow it. When I see a shooting star, I have a conscious visual experience that lasts for a brief period of time, then ends.

These alleged facts, if they really are facts, are a serious impediment to phenomenological temporalism. In physical spacetime, there is no such thing as absolute duration (apart from the zero-duration of point-events). Conscious experience therefore appears to have a property that it cannot have, if it exists in spacetime.

Second, there is the fact, or alleged fact, that each moment of conscious experience simultaneously exemplifies multiple phenomenal qualities, and absolutely so. When I see a book, I have a conscious experience that simultaneously has a certain phenomenal shape, a certain phenomenal size, and a certain phenomenal color. This simultaneity is absolute, in the sense that the experience instantiates these properties simultaneously according to every complete and accurate description of the world. Under no correct description do I see the book as having size only before or after I see it as having color. When I hear a violin accompanying an oboe, I have a conscious auditory experience that is simultaneously characterized by string and woodwind phenomenology; this simultaneity is also absolute: there is no acceptable redescription of my experience according to which it presents the instruments as taking turns.

These alleged facts, if they really are facts, also challenge the idea that conscious experience exists in the same time as physical events. In physical spacetime, absolute simultaneity entails spatial and temporal coincidence. An experience consisting of an absolutely simultaneous instantiation of various qualia can therefore occur in spacetime only as a singular point-event. But as we shall see, it is hard to combine the idea that our most basic experiences are mere point-events with a plausible account of the relationship between mind and body.

Some think that we needn't particularly worry about apparent conflicts between the scientific understanding of spacetime and what we believe about our own conscious experience. When we attribute temporal features to experience, the temporal concepts we use are our everyday, pre-theoretical ones. These concepts evolved as cognitive mechanisms for coping with the challenges that face medium-sized inhabitants of medium-sized ecosystems. When we apply them to our own experience, the beliefs we thereby form are false, as are all beliefs that run counter to the scientific understanding of time. We retain such beliefs only because they serve us well in our struggle to survive and flourish. They are extremely useful and deeply-ingrained misrepresentations of reality.[7]

[7] See Savitt (2009), 357–8, Callender (2008a), and Barbour (1999), 28–38.

But this underestimates the challenge that consciousness poses for the scientific conception of time. The real challenge is to account for the fact that the scientific conception of time appears to be directly refuted by the evidence of introspection. When we look inward, we seem to find conscious states and processes having temporal features that cannot exist in physical spacetime. Even if evolutionary psychology explains why we believe that consciousness has these features, the fact remains that the relevant beliefs appear to be *verified* by various aspects of our own mental lives. It is this apparent *justification* of the beliefs by our introspective evidence that needs to be explained, or explained away, rather than the beliefs' practical role in everyday life.[8]

The fact that Russell and Poincaré would have taken the central thesis of this book seriously should already be enough to soften some incredulous stares. In Chapters 4 and 5, I shall argue that Russell and Poincaré were indeed correct to think that there are deep tensions between a temporalist conception of experience and a modern scientific view of time. My hope is that when you see how strange are the terms on which alone conscious experience can exist in physical time, you'll be more willing to entertain the suggestion that perhaps it does not exist in physical time; you will scrutinize the suggestion, rather than staring at it in blank disbelief.[9]

1.3 Terminology

Ordinary language tends to blur the distinction between what we might call the "objective" and "subjective" structure of experience. We describe both a quilt and a visual perception of it as multicolored, both a noise and an auditory perception of it as varying in tone. This kind of talk is fine for everyday purposes, but it can be misleading when we want to talk about the relationships *between* objective and subjective structure. Introducing a modest amount of jargon at this point will help us avoid confusion in later discussion.

By "physical time," "physical space," and "physical spacetime," I mean the time (space, spacetime) in which physical events take place. This usage does not exclude the possibility that there are non-physical events that occur in physical time or spacetime.

[8] I take it that this is what Einstein was getting at in the remarks reported by Carnap at the head of this chapter.

[9] The attentive reader will notice that I have already stated the purpose of Chapters 4 and 5 to be that of reconciling facts about phenomenal simultaneity (Chapter 4) and the stream of consciousness (Chapter 5) with an atemporal conception of experience. This points to a flaw in the book's narrative structure which I have allowed to stand only because I know of no way to remedy it without adding greatly to the length and redundancy of the overall discussion. I can only ask my readers to recall the double purpose of Chapters 4 and 5 when we come to them: these chapters are meant to motivate as well as defend the atemporal position.

I take it as a basic assumption that our best understanding of physical time is the understanding we find in mainstream contemporary physics. This is the understanding of time as an aspect of relativistic spacetime.[10]

It is true that even at the highest levels of scientific discussion, there are disagreements about the role of time, or our concept of it, in our best physical theories. Some go as far as to claim that time (or the concept of it) plays no indispensable role in our best theories. Those who take this view do not necessarily (or even typically) deny that time is real—that it really is the case that events occur in time, and stand to one another in various temporal relations. What they deny is that time is *fundamental,* holding instead that facts about physical time reduce to more basic physical facts expressible in purely atemporal terms. But the physical time they aim to reduce to something more basic is, on the contemporary mainstream view, an aspect of relativistic spacetime.[11]

Throughout the book, I use the phrase "experience as of X" to denote a conscious experience with the phenomenal properties (or "qualia") that are distinctive of the sort of experience that a human being typically has if he or she consciously perceives X. The "as of" idiom is just a convenient way of referring to the phenomenal properties of an experience, which has those properties logically independently of what is typical for actual human experiencers. Like any way of talking about the conscious properties of experience, the device has its limitations. The experience one has when subject to the waterfall illusion, for example, has a very distinctive phenomenal quality, but this quality resists description in terms of the "as of" idiom, or in any succinct way ("phenomenally upward-flowing-cliff-face-ish" is the best I can manage).

Experience "as of" is to be distinguished from experience "of." According to my way of speaking, you can have an experience *of* something only if that something exists; likewise, one may be said to "experience X" only if X exists. Experience *as of* is not subject to this restriction. An experience as of X need not be an experience of X, nor need an experience of X be an experience as of X. The experience you have when you see a mirage is an experience as of water without being an experience of water (there being no water present to experience), and if, due to the influence of some drug, you have a visual image of a giant toad when you look at your

10 While I take the correctness of the relativistic conception of time (or spacetime) as a basic assumption, I do address some of the more common objections to the relativistic conception in an Appendix at the end of the book.

11 For the idea that time might be reducible to some more fundamental physical phenomenon, see Barbour (1999), Rovelli (2008), and Callender (2010). Obviously, I too deny that time is fundamental, since I hold that it reduces to something broadly phenomenological. But I take no stand on the question of whether we should think of time as a *physical* fundamental, that is, as having fundamental status relative to the (on my view phenomenalistically reducible) physical domain.

waste basket, your experience is of, but not as of, a waste basket. In this case, we may also say that you perceive a waste basket, but not that you perceive a toad. Your perception in this case is an example of misperception. (So, as I use the term "perception," not all perception is veridical: misperception counts as a species of perception.)[12]

Given my usage of the "as of" idiom, there is no such thing for an ordinary human being as an experience as of a million-sided polygon. This is not because an ordinary human being cannot *perceive* a million-sided polygon (he can), nor because there is no phenomenology associated with ordinary human viewings of million-sided polygons (there is). It is because there is nothing phenomenally distinctive about ordinary human perceptions of million-sided polygons—nothing that phenomenally distinguishes a typical human perception of a million-sided polygon from a typical human perception of a 900,000-sided polygon, or a million-and-one-sided polygon, or a circle. By contrast, there is such a thing for us as an experience as of a square, since there are qualia that are distinctive of the sort of experience that an ordinary human being typically has if he consciously perceives a square.

For ease of exposition, I also use the phrase "appearance of X" as synonymous with "experience as of X." So, for example, if I say that an appearance of duration need not be an enduring experience, what I mean is that an experience as of an enduring thing need not be an experience that itself endures.

In addition, I employ the standard device of referring to various phenomenal features of experience with phrases like "phenomenal redness," "phenomenal loudness," and "phenomenal duration," these being the qualia by virtue of instantiating which an experience is an appearance of (respectively) something red, loud, or enduring.

I often speak of qualia and qualia-instantiations. A quale is a property such that for an experience to have it is for there to be a certain something it is like to have that experience. Starting out, I make no assumptions about the ontological

[12] My preferred usage of the "as of" idiom was fairly common in mid-twentieth century philosophy of mind; see, for example, Smart (1963), 19–20, 23–24. More recently, some writers have taken to using "experience as of X" to imply something about how the experiencer conceives of what he experiences (or takes himself to be experiencing); see, for example, Dretske (2003), 70–71. On this alternative usage, to say that someone has an "experience as of water" is to say, among other things, that he conceives of that which he perceives (or takes himself to perceive) as water. (The conceiving needn't be active or voluntary; it might be passive and automatic; see McDowell (1994), 10–11.) Thus, on this usage, there might be a difference between an experience as of water and an experience as of H_2O, or an experience as of Clark Kent and an experience as of Superman. On the older usage which I follow in this book, there is no difference between an experience as of water and an experience as of H_2O. More generally, in speaking of an experience as of X, we do not, in my usage, imply anything about which concepts (if any) the experiencer deploys in connection with the experience.

status of qualia. In particular, I leave it open whether qualia are properties that a thing might have by satisfying certain purely physical conditions. As I use the term "qualia," a reductive physicalist agrees that experiences instantiate qualia: he just maintains that for a quale to be instantiated is for some purely physical (e.g. neurochemical) state of affairs to obtain.

A quale is a property. I take no stand here on the ontological status of properties, for example, on whether they should be understood as Platonic universals, or as classes of possibilia, or as mereological wholes, etc. I do assume that not all properties, and not all qualia, are actually instantiated. I want to stay neutral on the question of what it is for a quale (or any other property) to be instantiated. In particular, I do not assume that there must be particulars over and above properties, such that for a property to be instantiated is for some particular to have the property. On some views, property-instantiation is not a relation between a property and an irreducible particular, but a relation between or among properties (so we might think of a particular object as a "bundle" of properties). I do not wish to take a side in this debate.

1.4 Introspection

I base many of the claims that I make in this book on introspection. For example, I claim that introspection frequently reveals to me experiences as of things changing, and I say that this gives me every reason to think that I actually have such experiences. I also claim that I never introspect an experience as of a ten-dimensional solid, or as of a million-sided polygon, or as of one sound lasting a nanosecond longer than another, and I say that this gives me every reason to think that I never have an experience as of a ten-dimensional solid, or a million-sided polygon, or one sound outlasting another by one nanosecond.

When it comes to introspection, my basic position is that it is a reliable source of information about one's own conscious experiences, provided that one undertakes it with due care, and bearing in mind certain natural limitations to which it is subject. Below, I give a more detailed statement of this basic position. But first, let me say a few words about what my position is *not*.

First of all, I am concerned only with our human experience. Whether insects have experiences, and if they do, whether and how they introspect them, does not concern me.

Second, I hold that introspection yields information only about individual experiences, *not* extended streams of consciousness, whole mental lives, or other large-scale phenomenological structures. The only information that a given instance of introspection can yield is information about an individual conscious experience, which is normally an experience as of the contents of some quite brief time interval (no more than a second). A single experience might contain other experiences as constituents; introspection of such an experience might reveal something about multiple experiences (including, perhaps, something about the relations among the simple experiences that constitute the complex one). But a single act of introspection cannot yield information about different experiences that are *not* constituents of a single, complex experience. If introspection reveals to me the existence of two experiences, those experiences must belong to a single experience: an experience by virtue of having which there is something it is like for me to introspect myself as having the two experiences.[13]

Third, I do not claim that we actually introspect all of our experiences. It seems to me that the great majority of my experiences are *not* introspected. When I walk down the street, I have many conscious experiences, but usually I pay no attention to them, focusing rather on traffic, crosswalks, and other pedestrians. We are not omniscient about our own experiences, if only because we do not form beliefs about all of our experiences.

Fourth, I do not claim that introspection can yield information about anything besides the *phenomenal* features of conscious experience. Through introspection, I can learn that I have an experience characterized by various qualia (e.g. visual and auditory qualia). But I do not claim that introspection on its own can tell me about the non-phenomenal features of an experience, such as its causes or effects (supposing the experience has causes or effects), the nature (or existence) of its neural correlate, its resemblance to experiences had on other occasions, etc. From a single act of introspection, I can gather information about the phenomenal character of a single (possibly complex) experience of mine; if introspection can give me more information than this, that is more than I claim for it here.

I now explain my position on introspection in detail. The position can be summarized as a principle, which I'll call the *Introspection Principle*. There are actually two ways to formulate the principle, depending on how we individuate conscious experiences with reference to their phenomenology.

On a broad individuation, it can be literally true that an experience could have had different qualia from those it actually has. Of course, there are limits to how different a conscious experience could have been while remaining the same experience. It would be strange, to say the least, to suggest that a purely visual experience could have been a purely auditory experience. But we are comfortable with saying things like, "This pain would have been even worse, if I hadn't taken an

[13] The basic point is due to Hurley (1998), 164–5; see also Bayne and Chalmers (2010), 510–11, 527–9.

aspirin." Also, and more important for present purposes, we do not usually balk at the idea that one could have introspected an experience that one did not, in fact, introspect, even though the act of introspection might have made some difference to the quality of the experience introspected. On a broad phenomenological individuation of experiences, it can be literally true that an experience that is not actually introspected could have been introspected, even if introspecting it would have made some difference to its phenomenal character.

If we are happy to work with a broad individuation of experiences, we can state the Introspection Principle as follows:

> Every experience that I have, *e*, is such that for each of its non-confounding qualia, Q, (1) if *e* has Q and I introspect *e* while cognitively unimpaired, and with a view to determining whether *e* has Q, and having taken all reasonable measures to ensure that my determination is correct, then I introspect that *e* has Q, and, (2) if *e* does *not* have Q, and I introspect *e* while cognitively unimpaired, and with a view to determining whether *e* has Q, and having taken all reasonable measures to ensure that my determination is correct, then I introspect that *e* doesn't have Q.

As natural as a broad phenomenological individuation of experiences may seem, there are reasons to prefer a narrow individuation of experiences, according to which it is essential to every experience that it has all and only the qualia it actually has. After all, if we are going to individuate experiences phenomenologically, we must in so doing assign paramount—indeed, exclusive—importance to experiences' qualia. It is hard to see how anything short of a narrow individuation of experiences is consistent with this.

But even if we individuate experiences narrowly, we should recognize that a sentence like, "If I had been paying attention to my own conscious experience when I saw the shooting star, I would have noticed that it was an experience as of a pale streak of light" points to *some* truth. If we don't want to say that the sentence is literally true, we can say that it points to the truth that if I had been paying attention to my own conscious experience, I would have noticed that I had an experience as of a pale streak of light—without any presumption that this experience would have been numerically identical to the experience I actually had.

To accommodate those who (in my view rightly) prefer to individuate conscious experiences narrowly, we can state the Introspection Principle as follows:

> For every experience I have, *e*, there is a possible experience, *e′*, that is just like *e*, with the possible exception of lacking whatever confounding qualia *e* possesses, such that for all qualia, Q, (1) if I had *e′*, and *e′* had Q, and I introspected *e′* while cognitively unimpaired, and with a view to determining whether *e′* had Q, and having taken all reasonable measures to ensure that my determination was correct, then I would introspect that *e′* had Q, and, (2) if I had *e′*, and *e′* did *not* have Q, and I introspected *e′* while cognitively unimpaired, and

with a view to determining whether *e′* had *Q*, and having taken all reasonable measures to ensure that my determination was correct, then I would introspect that *e′* did not have *Q*.

Both versions of the Introspection Principle refer to "confounding qualia" and "cognitive impairment." What are these?

Cognitive impairment includes things like mental malfunction (due to fatigue, intoxication, etc.), inattention (due to laziness or carelessness), and psychological disturbance or disruption (due to being startled or frightened, for example). Cognitive impairment often diminishes the reliability of ordinary sense-perception, and there is no reason to think that it can't have the same effect on introspection.

Suppose that I am not suffering from any cognitive impairment. And suppose that in this unimpaired state, I introspect an experience, *e*, with a view to determining whether it has a certain phenomenal property, *Q*, and having taken all reasonable measures to ensure that my determination is correct. Then I am, as I shall say, *introspecting e under ideal circumstances, relative to Q*.

With this bit of jargon in hand, I can explain what I mean by "confounding qualia." I say that *Q* is a *confounding quale* just in case it is impossible to introspect any experience under ideal circumstances, relative to *Q*.

The most extreme example of a confounding quale is what we might call "phenomenal obliviousness." This is a phenomenal property by virtue of instantiating which an experience is, among other things, an experience as of some state of affairs obtaining in the absence of any introspection. If there is such a phenomenal property, it is hard to see how its instantiation could ever be introspected; an instantiation of phenomenal obliviousness would be essentially unintrospected, in the same way that an instantiation of unobservedness is essentially unobserved. We could think of it as a phenomenological blindspot.[14]

Phenomenal obliviousness is one confounding quale, but it may not be the only one. There might be qualia that render any experience that has them immune to *ideal* introspection, without rendering an experience completely immune to introspection (as phenomenal obliviousness does). For example, maybe introspecting an experience under ideal circumstances necessarily involves a certain kind of focus, attention, or clarity of mind, and maybe there is such a thing as an experience as of things existing in the absence of these factors. So we might speak of "phenomenal carelessness," in contrast to phenomenal obliviousness. Assuming that there is such a property as phenomenal carelessness, it is impossible to introspect any experience that instantiates it under ideal circumstances, even

14 Or as an "elusive phenomenology," as Uriah Kriegel puts it. See Kriegel (2013), 1171.

though it would be possible to introspect that an experience was phenomenally careless, under circumstances that were not ideal.[15]

There are qualia that do not characterize any experience that you or I or any other human being can have. That means that no human being can introspect the instantiation of such a quale, but it does not mean that such a quale is confounding. No doubt, there could exist a being that had experiences as of ten-dimensional objects, million-sided polygons, etc. There is no reason to think that such experiences would have to be attended by cognitive impairment, or that they would have to be experiences as of things existing in the absence of any introspection, etc. Phenomenal ten-dimensionality and phenomenal million-sidedness are not confounding qualia.

The Introspection Principle basically says that the non-confounding phenomenal features of my conscious experiences are available to ideal introspection (by me). Let's briefly consider how this supports the kind of introspection-based claims that I mentioned at the beginning of this section.

I said that I often introspect experiences as of change, and that this gives me every reason to believe that I do have such experiences. For example, I currently introspect an auditory experience of a swelling orchestra. I am cognitively unimpaired. I have heard this sound many times before, and on many occasions I have subjected the auditory experiences it gives me to introspective scrutiny. I have considered the suggestion that such an experience has no property of phenomenal change—that phenomenal change is in fact an illusion that somehow arises from combinations of phenomenally changeless experiences—and introspected upon my experience with a view to determining whether it is consistent with this suggestion. In short, I introspect my auditory experience under circumstances that are ideal relative to the property of phenomenal change, and what I find is a phenomenally changing auditory experience, rather than an auditory experience that lacks the property of phenomenal change. Clause (2) of the Introspection Principle entitles me to conclude that I do, in fact, have an experience as of change.

I also said that I never introspect an experience as of a ten-dimensional solid, and that this fact gives me every reason to think that I never have an experience as

[15] Could there be an experience that had *only* confounding qualia? It's hard to see how. If I have an experience as of a state of affairs existing in the absence of introspection, my experience is still an experience as of some state of affairs, and so presumably has some more or less ordinary qualia answering to the apparent shapes, sizes, colors, motions, sounds, etc. of the thing or things involved in the apparent state of affairs. And no matter how tired, inattentive, or distracted I am, my experience has some more or less ordinary qualia corresponding to various features of the things of which I am dimly aware (or that distract me), or qualia characteristic of basic bodily sensations. The closer one comes to imagining a state of *pure* absent-mindedness or mental incapacitation unadulterated by any bodily or perceptual phenomenology, the closer one comes to imagining a state of unconsciousness.

of a ten-dimensional solid. Well, suppose that you sell me some pills that you say will allow me to have experiences as of ten-dimensional solids. I take one of the pills, and wait for the fun to start. But it never does. The pill has no effect at all. Try as I might, I cannot introspect any experience as of a ten-dimensional object (or even as of a four-dimensional object)—this, despite the fact that I am not suffering from any cognitive impairment, and have taken all reasonable measures to ensure that I do introspect such an experience, should I have one. Clause (1) of the Introspection Principle entitles me to conclude that I do not have an experience as of a ten-dimensional solid.

Now suppose that the pills you gave me were supposed to be painkillers rather than hyperdimensional hallucinogens. I have just had my wisdom teeth extracted, and am in considerable pain, though not so much as to induce any significant cognitive impairment. I take one of the pills. But the pill has no effect. I double the dose. Still no effect. When I introspect with a view to finding evidence of relief, I find no such thing; all I find is an experience as of a throbbing pain in my jaw. Clause (2) of the Introspection Principle entitles me to conclude that I am indeed having a painful experience.

To many people, including me, there is something highly suspect about the idea of a conscious experience that is completely immune to introspection by the person who has it. But it is important not to be dogmatic about this. As we have seen, there is some basis for thinking that your mental life includes qualia-instantiations that are immune to introspection by you, namely, instantiations of phenomenal obliviousness. It would be a mistake to insist dogmatically that *all* the phenomenal features of your mental life are available to introspection by you, just as it would be a mistake to insist that all the non-microscopic features of a strawberry are observable (if the strawberry is unobserved, it has a non-microscopic property—that of being unobserved—that is immune to observation).

But what about qualia other than phenomenal obliviousness? Let's call such qualia "ordinary qualia" (even though the experiences they characterize might be quite extraordinary in terms of their what-it-is-likeness). There is nothing in the logic of an ordinary quale to prevent you from introspecting an instantiation of it in your mental life. It is impossible to introspect an instantiation of phenomenal obliviousness, for the same reason as it is impossible to observe an instantiation of unobservedness. But mustn't it be possible, at least in principle, to introspect the instantiation of any ordinary phenomenal property that characterizes your experience, just as it is possible, at least in principle, to observe the instantiation of any ordinary non-microscopic feature of a strawberry (that is, any non-microscopic property besides the property of being unobserved)?

Some take it to be an a priori truth that the answer to this question is Yes. I confess to being among them. But there are those who suggest that you could have an experience characterized by some ordinary quale, even if it were impossible for you ever to introspect an instantiation of that quale.

I am going to resist the temptation to accuse those who make this suggestion of failing to take consciousness seriously. Instead of taking this as an occasion for criticism, I shall take it as an opportunity for diagnosis.

Why would anyone think that his experience could have an ordinary conscious property that he was utterly incapable of introspecting (i.e. a quale such that he is incapable of introspecting an instantiation of that quale)? Well, suppose that you are watching a strobe light. At first, the light flashes rather slowly, but as you watch, the rate of flashing increases. Up to a certain point, introspection reveals experiences as of flashing light. But at some point, as the strobe light flashes more and more rapidly, all that you find yourself able to introspect is an experience as of a single, steady point of light. That is all you can introspect, even if your circumstances are ideal for the purposes of introspection. It's not that the light source is no longer flashing. It is. It's just that it is flashing at such a high rate that you can no longer discriminate the individual flashes from one another.

Call this the Strobe Case.

Now, you might wonder about the experiences that took place towards the end of the Strobe Case. You might conjecture that those experiences really *were* experiences as of a flashing light, despite the fact that your best introspective efforts revealed only experiences as of a steadily shining light. Maybe there were visual experiences as of flashing throughout the Strobe Case, despite the fact that beyond a certain point, it was a phenomenal flashing that was utterly beyond your powers to introspect. It's true that at a certain point, introspection did clearly reveal the presence of an experience as of a steadily shining light, but maybe this experience was, somehow, really an experience as of a very rapidly flashing light.

I do not see that we have any reason to think that there is any experience as of flashing light towards the end of the Strobe Case. But at the same time, there is nothing in the foregoing conjecture that conflicts with the Introspection Principle, or with the view that any ordinary quale that characterizes your experience is a quale whose instantiation you are capable of introspecting.

The suggestion was that in the Strobe Case, there was a sequence of instantiations of phenomenal flashing, the first of which you could easily introspect, and the last of which you could not have introspected no matter how hard you tried. If we think of all these qualia-instantiations as arising from chemical activity in your brain, we might say that your brain generates a sequence of appearances of flashing light, where the phenomenal flashing of the earlier appearances is introspectively accessible to you, but the phenomenal flashing of the later ones is not.

This suggestion conflicts with the claim that you have introspective access to all of your ordinary qualia only on the further assumption that the later instances of phenomenal flashing are part of *your* conscious mental life.

But the very fact that the later phenomenal flashing (or alleged flashing) is utterly immune to introspection by you is a compelling reason to deny that it is part of your conscious mental life. Let me explain.

Following Ned Block, we can imagine a situation in which we equip every resident of China with a telegraph and a set of instructions. Each set of instructions will be different from the others, but they will all look something like this: "When you receive three dots from X, send six to Y"; or this: "When you receive eleven dots from X within ten seconds of receiving two dots from Y, send four dots to Z"; etc.[16]

Assuming that there are as many people in China as there are neurons in a human brain, we can imagine that by following their instructions, the Chinese collectively engage in behavior that parallels the synaptic signalling that takes place in the brain of a blind person when he has ordinary auditory experiences. For example, we can imagine that a certain rapid sequence of telegraphic dots in the Chinese set-up corresponds to the synapse-firings that occur in George Shearing's brain when he hears middle C on the piano, and that a slower sequence of telegraphic dots in the Chinese set-up corresponds to the synaptic activity that occurs in Shearing's brain when he hears C-sharp.

Arguably, in the scenario just described, the activity of the Chinese results in auditory experiences just like those that Shearing has when playing the piano. Anyway, let's suppose that the telegraphic activity does result in such experiences. Even so, the phenomenal character of the musical experience owes nothing to the phenomenal characters of the telegraphers' mental states. We can imagine that all of the telegraphing is taking place in silence, or that everyone involved in the experiment is deaf.

Taking this thought experiment a step farther, suppose we grant, speculatively, that conscious mental events are capable of interacting with or depending on one another in some causal or quasi-causal way. We can then entertain a situation in which various conscious experiences occur in a pattern paralleling the pattern of telegraphic activity described above, and therefore also paralleling the neural activity of Shearing's brain. For instance, we can imagine a cascade of visual sensations that unfolds in parallel with developments in the Chinese telegraph experiment (replacing telegraphic dots with phenomenal points of light) and therefore also parallel to the activity in Shearing's brain. Call this hypothetical

16 See Block (2007), 71–3.

system of visual sensations the "Visual Brain," and give the name "Victor" to the conscious subject whose brain the Visual Brain is.

The phenomenal characters of the sensations that make up the Visual Brain—its "visual neurons," if you will—are irrelevant to the phenomenal character of Victor's auditory experiences. In the case described, the visual experiences do not even belong to the same conscious mental life as the auditory experiences, this being the mental life of a person who is blind from birth. It is as if one were to take a standard reductive analysis of a conscious experience into neurochemical states, and then postulate that each of these states has qualia of its own, although not necessarily qualia that characterize the consciousness of the creature in whose brain the neurochemical states occur. The visual experiences that make up the Visual Brain clearly have an important bearing on the auditory phenomenology of Victor's mental life, but they are not part of that life in the way that the auditory phenomenology is—they are not *Victor's* experiences, any more than the experiences of billions of homuncular Chinese would be *my* experiences, if my brain were composed of homuncular Chinese instead of ordinary neurons.

Now suppose that Victor's blindness is cured. To his delight, he can now enjoy paintings, sunsets, and the night sky just like any sighted person. He sits on a sand dune, watching the Sun sink into the sea. He drinks in the delicate pinks and blues of the evening sky. Night falls. Venus appears, a steady point of light outshining neighboring stars.

Despite this radical enrichment of Victor's phenomenology, the basic structure of his brain, the Visual Brain, has remained the same—it remains a brain composed of visual sensations. In particular, we may suppose that the experience Victor has as he gazes at the Evening Star has for its neural correlate an experience consisting of a very rapid sequence of visual sensations as of shining light, paralleling some rapid sequence of dots in a suitably updated telegraph network, and some rapid sequence of synapse-firings in the brain of a sighted version of George Shearing. Of course, Victor is unaware of the events in this rapid sequence, just as he is unaware of the events in the sequences that constitute his auditory experiences, and just as an ordinary human being is unaware of the opening and closing of ion-channels in the cells of his own brain. Victor's other visual experiences have other neural correlates in the Visual Brain; for example, we may suppose that when Victor introspects a visual appearance of a blinking light, the neural correlate of the experience he introspects is a slower sequence of sensations within the Visual Brain.

Let's return to the Strobe Case. We were considering the suggestion that in that case, there occurred experiences that were completely inaccessible to introspection by you (certain experiences as of very rapid flashes of light). I conceded that it was possible, at least logically, that such experiences occurred in that case, but asserted that if they did, they were not among *your* experiences. We can now see the rationale for this assertion.

If, in the Strobe Case, there occurred experiences as of flashes of light that were unintrospectible by you, those experiences related to your mental life in much the same way as the flashing experience that serves as the neural correlate of Victor's perception of Venus relates to Victor's mental life. The flashing experience that occurs as part of the Visual Brain is introspectively inaccessible to Victor; the flashing experiences that supposedly occur towards the end of the Strobe Case are introspectively inaccessible to you. Do the unintrospectible-by-you experiences that supposedly occur in the Strobe Case have a bearing on your behavior? The experiences that constitute the Visual Brain have a bearing on Victor's behavior. Do the unintrospectible-by-you experiences that supposedly occur in the Strobe Case make a difference to your introspectible visual phenomenology? The experiences that constitute the Visual Brain make a difference to Victor's introspectible phenomenology. Would you have had an introspectible experience as of flashing light at the end of the Strobe Case if there had occurred in you a slower sequence of light-experiences? Victor would have had an introspectible experience as of flashing light if there had occurred a slower sequence of light-experiences in the Visual Brain.

The visual sensations that constitute Victor's brain are not part of Victor's conscious mental life. Why not? They make a difference to Victor's behavior. They even make a difference to the phenomenal quality of Victor's visual experience. The visual sensations that constitute Victor's brain are not part of Victor's conscious mental life because it is impossible for Victor to introspect them.

By the same token, the fact that it is impossible for *you* to introspect the flashing light experiences that supposedly occur towards the end of the Strobe Light Case implies that those experiences are not part of your conscious mental life.

The point generalizes. Given *any* ordinary quale—any phenomenal property other than the pathological property of phenomenal obliviousness—that quale characterizes an experience in a given subject's conscious mental life only if it is a quale whose instantiation that subject can introspect (even if he doesn't actually introspect it).

I started by asking why some people think that they might have experiences with ordinary conscious properties that they are utterly incapable of introspecting. My explanation—the diagnosis promised earlier—is that these people are conflating two alleged states of affairs, one possible (though far-fetched), the other impossible. The possible state of affairs is that in which a subject's brain produces experiences (e.g. of extremely rapid flashes) that are introspectively inaccessible to the subject. The impossible state of affairs is that in which a subject has experiences that are introspectively inaccessible to the subject. Failing to distinguish between these states of affairs (or alleged states of affairs) can lead us to mistake an impossibility for a possibility.

I have asserted that introspection is a reliable source of information about conscious experience, provided that we employ it with due care, and I have offered some positive reasons to think that the phenomenology that characterizes a person's mental life is, by and large, introspectively accessible to that person. It remains to consider the arguments that have been offered *against* the reliability of introspection.

Some of these have been around for a long time; it seems to me that Frank Jackson has effectively countered them.[17] But a new family of arguments critical of introspection has lately emerged. These arguments are well represented in the writings of Eric Schwitzgebel.

According to Schwitzgebel, "the introspection of current conscious experience is . . . highly untrustworthy," to such an extent that "we make gross, enduring mistakes about even the most basic features of our currently ongoing conscious experience . . . even in favorable circumstances of careful reflection, with distressing regularity."[18]

He gives three kinds of argument for this pessimistic view.

First are what I'll call the "straw man" arguments. These are a recurring theme of Schwitzgebel's discussion; here is a representative sample:

> Now, you're a philosopher, or a psychologist, presumably interested in introspection and consciousness and the like . . . You've had emotional experiences, and you've thought about them, reflected on how they feel as they've been ongoing or in the cooling moments as they fade. If such experiences are introspectible, and if introspection is the diamond clockwork often supposed, then you have some insight. So tell me: Are emotional states like joy, anger, and fear always felt phenomenally—that is, as part of one's stream of conscious experience—or only sometimes? Is their phenomenology, their experiential character, always more or less the same, or does it differ widely from case to case? For example, is joy sometimes in the head, sometimes more visceral, sometimes a thrill, sometimes an expansiveness—or, instead, does joy have a single, consistent core, a distinctive, identifiable, unique experiential character?

These are all interesting questions, but they have no obvious bearing on the reliability of introspection. Introspection alone cannot possibly tell me whether all my states of joy have a phenomenal character in common, for the simple reason

[17] In Jackson (1973). [18] Schwitzgebel (2008), 247.

that I cannot introspect all my joyful states together. To answer the questions that Schwitzgebel asks here, I must, at a minimum, draw on the resources of memory (and probably other cognitive capacities as well, such as inference and analysis). If I am unsure how to answer these questions (or the many similar ones that Schwitzgebel poses in this connection: "Do you really enjoy Christmas?", "Are you happier weeding or going to a restaurant with your family?", etc.), the most that this shows is that a combination of introspection and other cognitive functions is fallible. In particular, the claim that judgments about whole *kinds* of phenomenal experience are fallible is consistent with *introspection* being entirely infallible, and certainly consistent with the circumscribed reliability that I have claimed for introspection.

The second class of arguments that Schwitzgebel offers against the reliability of introspection are what I'll call the "zombic" arguments. For example:

> [I]f malevolent neurosurgeons from Alpha Centauri may massage and stoke our brains, I see no reason to deny them the power to produce directly the judgment that one is having reddish phenomenology, while suppressing the reddish phenomenology itself. Is this so patently impossible?

It's not impossible at all—or so let us grant. In fact, let's grant that the neurosurgeons could have suppressed *all* of my phenomenology, without suppressing any of my actual judgments, including the judgments in which I attribute various conscious experiences to myself. Let us grant that my zombie twin judges himself (erroneously) to have various conscious experiences. The zombie behaves just like I do: he is no less sincere than me when describing what he mistakenly believes to be his experiences, he is no less earnest when asserting that the soup tastes delicious, etc.

None of this poses any threat to the reliability of introspection.

There is a possible world in which a handless counterpart of me mistakenly judges himself to have hands; this does not cast doubt on my own judgment that I have hands. No more does the existence of a possible world in which a zombified counterpart of me mistakenly judges himself to have conscious experiences cast doubt on my own belief that I am conscious. I know I have conscious experiences because in addition to offering sincere descriptions of, and earnest assertions about, my conscious experiences, I am introspectively aware of my experiences. If Schwitzgebel wishes to dispute this, he must find a new argument.[19]

[19] Anyway, if the possibility of Schwitzgebel's Alpha Centaurians casts doubt on our introspective judgments, it is only a generic (and arguably bogus) skeptical doubt that we can equally cast on almost any judgment, including our ordinary perceptual judgments.

The third class of arguments that Schwitzgebel offers against the reliability of introspection are what I'll call the "interpersonal disagreement" arguments. These have the following form:

ID1. There are people who have experiences that are identical in a given phenomenal respect, but who disagree about the phenomenal character of their experiences in this respect, even upon ideal introspection.
ID2. If ID1 is true, then some people mistake the phenomenal characters of some of their experiences, even upon ideal introspection.
ID3. So, ideal introspection is fallible.

The second premise seems uncontroversial. But why should we accept the first? Here is Schwitzgebel:

> Look around a bit. Consider your visual experience as you do this. Does it seem to have a center and a periphery, differing somehow in clarity, precision of shape and color, richness of detail? Yes? It seems that way to me, too. Now consider this: How broad is that field of clarity? Thirty degrees? More? Maybe you're looking at your desk, as I am. Does it seem that a fairly wide swath of the desk—a square foot?—presents itself to you clearly in experience at any one moment, with the shapes, colors, textures all sharply defined? Most people endorse something like this view when I ask them.[20]

Schwitzgebel then reports that the same people who judge their regions of visual sharpness to be about thirty degrees of their visual fields can subsequently be brought to judge that the region of sharpness is actually much narrower—just a degree or two of arc around the center of the visual field. They are brought around to this view by careful introspection of their visual experiences under controlled conditions (of slowly moving a playing card from peripheral to central vision, etc.), and having been told that the region of sharpness is probably much narrower than they initially judged:

> Most of the people I've spoken to, who attempt these exercises, eventually conclude to their surprise that their experience of clarity decreases substantially even a few degrees from center. Through more careful and thoughtful introspection, they seem to discover—in fact, I think they really do discover—that visual experience does not consist of a broad, stable field, flush with precise detail, hazy only at the borders. They discover that, instead, the center of clarity is tiny, shifting rapidly around a rather indistinct background. My interlocutors—most of them—confess to error in having originally thought otherwise.[21]

Schwitzgebel continues:

> If I'm right about this, then most naive introspectors are badly mistaken about their visual phenomenology when they first reflect on it, when they aren't warned and coached against

[20] Schwitzgebel (2008), 254. [21] Schwitzgebel (2008), 255–6.

a certain sort of error, even though they may be patiently considering that experience as it occurs.[22]

So far, there is nothing in this case to cast doubt on the Introspection Principle stated earlier. The initial introspective judgments about the focal area of the visual field are not made under ideal circumstances, since the people making the judgments have not taken all reasonable measures to ensure that their introspective judgments about these experiences are correct (e.g. they have not taken various measures that would have "warned them against a certain sort of error"). The fact, assuming it is a fact, that these judgments were mistaken therefore does not conflict with the claim that introspection is reliable under ideal circumstances.

At this point, Schwitzgebel imagines someone who insists that his region of clarity really is about thirty degrees, even in the light of ideal introspection. Schwitzgebel does not report that any actual person who introspects under ideal conditions insists on this, and it is hard to see how merely *imagining* such a person lends any credence to ID1, which claims that there actually are people like this. But let us suppose that some of the people involved in the experiments Schwitzgebel describes actually do insist that the region of clarity of their visual fields is about thirty degrees, even upon ideal introspection (having been coached and warned against a certain sort of error, etc.).

This by itself does not support ID1. To get support for ID1, we have to suppose further that the people who (upon ideal introspection) judge their focal areas to be about thirty degrees have the same quality of visual experience (as regards phenomenal region of clarity) as those who (upon ideal introspection) judge their focal areas to be only a couple of degrees.

What is Schwitzgebel's argument for the latter claim? Why does he think that in the case described the subjects who ascribe different areas of phenomenal focus to their respective visual fields actually have visual fields with focal regions of more or less the same size? After all, if two subjects report, in the light of ideal introspection, different qualities of visual experience, it is surely quite natural to conclude that the phenomenal characters of their visual experiences really *are* different.

Here is what Schwitzgebel has to say about this:

Human variability, though impressive, usually keeps to certain limits. Feet, for example—some are lean and bony, some fat and square, yet all show a common design: skin on the outside, stout bones at the heel, long bones running through the middle into toes, nerves and tendons arranged appropriately. Only in severe injury or mutation is it otherwise. Human livers may be larger or smaller, better or worse, but none is made of rubber or attached to the elbow. Human behavior is wonderfully various, yet we wager our lives daily on the predictability of drivers, and no one shows up to department meetings

[22] Schwitzgebel (2008), 256.

naked. Should phenomenology prove the exception by varying radically from person to person—some of us experiencing one hundred degrees of visual clarity, some only two degrees . . . with as little commonality as these diverse self-attributions seem to suggest?[23]

The first thing to say about this is that there is, actually, quite a lot of variation in the structure and function of feet and livers, particularly at the fine-grained cellular level. But even if this were not so, it would hardly support Schwitzgebel's claims about the homogeneousness of human experience. The human brain is the most complex and adaptable object in the known universe. This is reflected by the fact that approximately 84 percent of human genes are expressed in the human brain. A human being *is*, to a decent first approximation, his brain. Inferring that there is probably not much variation in brain function from the (alleged) fact that there is not much variation in foot function is like inferring that there's probably not much variation among pictures, since there's not much variation among picture frames.[24]

Furthermore, we presently know very little about the brain. Scientists were recently surprised to discover that neurons in higher mammals are arranged in three-dimensional rectangular grids: that's the level of understanding we're at. As for detailed knowledge of the low-level neural activity associated with particular forms of experience—that is the stuff of a hoped-for neuroscientific future. Certainly we are not in a position to say, of any two subjects who offer divergent introspective reports of their visual experiences, that the neural correlates of the experiences they report are identical in all relevant physical or functional respects.[25]

Indeed, we know that people often *do* have different qualities of experience, even when equipped with sense-organs that are, as far as we can tell, entirely similar in relevant structure and function. The same food can taste different to two people—sometimes very different, and often at least slightly different. When one person says that the chili tastes spicy and another says that it tastes bland, we don't conclude that one of them is making a mistake about his flavor-phenomenology. We conclude that they have different kinds of flavor experience. Two people with equally good vision, one a city-dweller, the other a country-dweller, might have different qualities of visual experience when walking together through the woods: the country-dweller might have experiences as of a rabbit in the underbrush, where the city-dweller just has experiences as of tangled twigs and leaves.

Phenomenological differences like these might come to light in the form of behavioral differences. But they might not, and in some cases it would be difficult

23 Schwitzgebel (2008), 264.

24 For the genetic claim, see Hawrylycz et al. (2012).

25 For the geometry of neural pathways, see Wedeen et al. (2012).

to find any behavioral sign of a difference in quality of experience. If, for example, the phenomenal coloration of one subject's visual field is systematically different from the coloration of another subject's visual field, this difference might never come to light in the subjects' behavior. Anyway, there *is* a behavioral difference between a subject who reports that his region of visual clarity is about thirty degrees and a subject who reports that his region of visual clarity is about three degrees: the former behaves by uttering the words: "about thirty degrees", whereas the latter behaves by uttering the words: "about three degrees."

But let's imagine a case in which we have the strongest possible grounds for supposing that two subjects have phenomenally identical visual experience. Well, not the strongest possible grounds, since that would involve the subjects being perfect physical duplicates, in which case, their behavior could never give us any reason to think that they disagreed in their introspective judgments (or in anything else). But we can certainly imagine two people, *A* and *B*, who have duplicate eyes. We can also imagine that the visual centers of their brains have been cross-wired in such a way that by toggling a switch, input from *A*'s eyes gets directed to *B*'s visual center, and input from *B*'s eyes gets directed to *A*'s visual center. We can imagine that toggling the switch does not have any effect on either subject's visual phenomenology (they can't tell based on introspection which position the switch is in, or whether it is being toggled back and forth). We can imagine furthermore that *A* and *B* have equal (and considerable) powers of introspective focus, concentration, etc., that they are both intelligent and rational, and that they are looking at the same scene (or scenes that perfectly duplicate one another).

Suppose that despite all this, *A*, upon ideal introspection, says: "I introspect my area of focus to be about three degrees," and *B* says (also upon ideal introspection): "I introspect my area of focus to be about thirty degrees."

If such a case were to arise, would it compel us to conclude that at least one of the subjects was making a false introspective judgment about the phenomenal character of his experience, despite introspecting the experience under ideal conditions?

No. Here are some alternative possibilities that are at least as plausible as the suggestion that *A* or *B* makes an introspective error:

(1) The subjects use phenomenal predicates (such as "sharp" and "blurry") to pick out different qualia, perhaps due to vagueness in the corresponding concepts (compare: "flat" and "non-flat") or to some other difference in phenomenological idiolect.

(2) The subjects understand different things by "introspect" (compare: "I heard a car" versus "I heard a car engine" versus "I heard the sound of a car engine").
(3) The subjects mean something different by "one degree of a visual field"; maybe by this *A* means one degree of the field that includes all his visual phenomenology (central, intermediate, and peripheral), while *B* means one degree of the part of his visual field to which he is attending in a certain way—a way in which he is not attending to most of his visual field.
(4) One of the subjects is lying, perhaps out of a stubborn refusal to admit openly that his former, naive introspective judgment was erroneous.
(5) *A* and *B* actually do differ in terms of their visual phenomenology: when *A*'s visual cortex receives input from *A*'s eyes or *B*'s eyes, the result is visual experience with a three degree field of sharpness; when *B*'s visual cortex receives input from *B*'s eyes or *A*'s eyes, the result is visual experience with a thirty degree field of sharpness.[26]

Now, it is not obvious that two subjects who satisfied all of the conditions described earlier (duplicate sense-organs, equal [and high] intelligence and rationality, indistinguishable visual stimuli, the toggling, etc.) might nevertheless describe their experiences in prima facie divergent terms ("three degrees" versus "thirty degrees"). But supposing that a situation were to arise in which two such subjects did describe their experiences in such terms, then any of the hypotheses enumerated above would be at least as plausible as the hypothesis that one of the subjects made a mistake in ideal introspection.

The basic problem with the interpersonal disagreement argument is that we have no good reason to accept its first premise (ID1), since the more reason we have to believe that some people have phenomenally identical experiences, the less reason we have to believe that those people make divergent introspective judgments about their experiences. If two subjects are perfect neural duplicates, then we have very good reasons to believe that they have the same phenomenal quality of experience, but no reason to think that they have different introspective beliefs about their experiences (since neural duplicates would also duplicate one another in all relevant publicly observable respects). If the subjects differ from

[26] What if we assume that *A* and *B* have duplicate visual cortices, with the position of the toggle switch determining whether the part of *A*'s brain that excludes *A*'s visual cortex receives input from *A*'s visual cortex or *B*'s visual cortex (and *mutatis mutandis* for *B*)? Then we would have to consider the possibility that when the part of *A*'s brain that excludes *A*'s visual cortex receives input from *A*'s visual cortex or *B*'s visual cortex, the result is visual experience with a three degree field of sharpness, and when the part of *B*'s brain that excludes *B*'s visual cortex receives input from *B*'s visual cortex or *A*'s visual cortex, the result is visual experience with a thirty degree field of sharpness.

one another in some neural respect, then they might offer prima facie divergent descriptions of their experience; but in such a case, the existence of the neural difference leaves open both (a) the possibility that the subjects' prima facie divergent descriptions are not expressions of divergent introspective beliefs [as in (1)–(4) above], and, (b) the possibility that the subjects' experiences actually differ in phenomenal character [as in (5)].

I conclude that we have good reasons to accept the Introspection Principle, and no good reasons to reject it. Recognizing that we can make mistakes about our conscious experiences when introspecting them under non-ideal circumstances should not diminish our confidence in the reliability of introspection under ideal circumstances. I shall therefore help myself to the deliverances of introspection, when introspection is undertaken under ideal circumstances, as it often enough can be.

1.5 A glance ahead

After this chapter, the book proceeds as follows.

Chapter 2: Space, time, and spacetime

Chapter 2 concerns the nature of physical time. We trace the Newtonian conception of time and space to its conceptual foundations, and see how the relativistic account of spacetime arises from its Newtonian origins. The chapter contains little that is new or controversial, and readers already familiar with (and sympathetic to) mainstream scientific views about spacetime may wish to gloss over it. My main goal in this chapter is to dispel the popular misconception that the relativistic account of spacetime is somehow inapplicable to the world of everyday experience, or discontinuous with our ordinary ways of thinking about time and space.[27]

Chapter 3: Phenomenal duration, succession, and change

The central thesis of Chapter 3 is that introspection gives us no reason to think that any aspect of our conscious experience is characterized by objective temporal succession: that introspection yields no evidence of successive sensations, or successive developments of a single sensation; no evidence of any sequential instantiation of qualia; etc. This pits me against a long tradition, upheld by philosophers as divergent in outlook as Kant and Mill, according to which we

[27] The purpose of Chapter 2 is to motivate and elucidate the relativistic conception of spacetime, not to defend it. I address some common objections to the relativistic conception in an Appendix at the end of the book.

have direct and incontrovertible evidence that conscious experience persists and changes over time. Breaking with this tradition overcomes the first objection to radical phenomenalism enumerated earlier.

Chapter 4: Phenomenal simultaneity

Define "serialism" as the view that qualia never get instantiated absolutely simultaneously, but only by whole temporal sequences of non-experiential events (like brain events). In Chapter 4, I argue that we have no purely introspective basis for thinking that serialism is false—and therefore no purely introspective basis for thinking that conscious experience is in any way characterized by simultaneity in physical time. This overcomes the second objection to radical phenomenalism enumerated earlier.

In Chapter 4, I also argue that the view that consciousness exists in physical time conflicts with certain facts related to phenomenal simultaneity, except on a dualist theory that accords conscious experiences spatiotemporal location, but no spatiotemporal extent. The only alternative to this dualist theory for someone who holds that consciousness exists in physical time is to affirm the possibility of a doubtfully coherent disunity of consciousness, in which, for example, someone simultaneously has visual and auditory experience, but no audiovisual experience.

A study of phenomenal simultaneity therefore not only fails to cast doubt on phenomenological atemporalism, but even forces a choice among atemporalism (à la Poincaré), pointillistic dualism (à la Russell), and a serialist metaphysics of consciousness that commits its proponents to a doubtfully coherent picture of the mind's relationship to itself.

Chapter 5: The stream of consciousness

In Chapter 5, I argue that radical phenomenalism is consistent with the relevant facts about the stream of consciousness. We compare temporalist and atemporalist accounts of the stream of consciousness, measuring different accounts by their ability to solve a central paradox connected with the stream of consciousness. The best temporalist account, due to Barry Dainton, takes streams of consciousness and their constituent experiences to be processes unfolding in physical time. But it turns out that an atemporal account of the stream of consciousness, modeled on Dainton's, solves the central paradox too.

Also in Chapter 5, I argue that any account of the stream of consciousness that assumes that conscious experiences occur in physical time must imply a failure of an otherwise universal natural symmetry (the "reversal invariance" of fundamental natural laws). The atemporal alternative to Dainton's theory avoids this difficulty. So in this respect, at least, it turns out that an atemporal account of the stream of consciousness is preferable to a temporal account.

Chapter 6: Idealism

Far from there being any compelling phenomenological argument in support of the idea that conscious experience has objective temporal qualities, we find, in the first five chapters of the book, that we actually have reasons to favor the atemporal view of consciousness, according to which conscious experience does not exist in spacetime at all. No doubt the atemporal view faces many challenges, but these are not, so to speak, internal to experience, having to do rather with how experience relates to the non-experiential. The phenomenological atemporalist owes us an account of how, in his view, conscious experience relates to the contents of spacetime, given that the relationship is neither spatial nor temporal.

If such an account is possible, it will have to take the form of a broadly idealist metaphysic, according to which spacetime and its contents logically supervene on facts about experience. In Chapter 6, we look at historical developments of this basic idea, the main two being: (1) the view that the physical facts reduce to facts about causal or quasi-causal potentialities for experience grounded either in minds or in supra-empirical "things-in-themselves," and (2) the phenomenalist view that the physical facts reduce to facts about fundamental phenomenological regularities. Also in Chapter 6, we survey the main challenges facing idealism in general, and phenomenalism in particular. This sets up the next three chapters, which are devoted to meeting these challenges.

Chapter 7: Phenomenalism: A first appraisal

Among idealistic metaphysics, phenomenalism is the most ambitious, and consequently the most heavily criticized. In Chapter 7, we address two of the main objections that apply to phenomenalism, but not to other, less ambitious idealist theories.

The first objection is that phenomenalism makes an unacceptable mystery of the fact that our conscious experiences have the order and coherence they do (or any order and coherence at all), since phenomenalists cannot call on any extra-experiential *causes* of experiences to explain this order and coherence; to this, I reply that the phenomenalist may regard the order and coherence of experience as a fundamental regularity, rather than a regularity explained by some underlying non-experiential posit.

The second objection is that phenomenalism forces us to say that perception never gives us access to anything but our own conscious experiences; to this, I reply that it is simply a mistake to think that phenomenalism requires us to say any such thing. In fact, a phenomenalist can even be a direct realist about perception, holding that the immediate objects of awareness in sense-perception are physical objects and events.

A further purpose of Chapter 7 is to argue that phenomenalism, unlike the idealism of Berkeley or Leibniz, does not commit its proponents to a conception of physical things as dependent for their existence on the existence of minds.

The moral of the chapter is that if we are going to be idealists, we ought to be phenomenalists.

Chapter 8: Radical phenomenalism

Chapter 8 advances the central theory of this book—a radical phenomenalism according to which all facts about spacetime and its contents reduce to broadly phenomenological facts. The theory is Leibnizian, inasmuch as it aspires to idealize facts about the temporal, and not just facts about the spatial. It also follows Leibniz in analyzing individual spatiotemporal states of affairs in terms of totalities of complete mental lives. However, unlike Leibniz, the theory I propose does not assume that these mental lives arise from independently given minds (or monads), or even that they enjoy more than a purely hypothetical existence. In these respects, the theory is closer to Mill's than Leibniz's.

Very roughly, radical phenomenalism equates a physical thing with a primitive tendency for increases in the total amount of conscious experience to correspond to increases in the probability that there are experiences interpretable as perceptions of that thing, with the probability going to 100 percent as the quantity of experience goes to infinity. The goal of Chapter 8 is to develop this rough proposal into a proper theory. Mainly this involves coming to an understanding of "increases in the amount of experience" that is consistent with an atemporal conception of experience, and captures the basic phenomenalist intuition that the physical world consists of whatever we would take ourselves to perceive, were we to have experiences as of exploring all of time and space.

Chapter 9: Idealism vindicated

The most important challenges to idealism in general are: (1) the problem of *unusual origins*: no amount of phenomenological potential is sufficient for the existence of ordinary physical things, since we can imagine scenarios (evil demons, envatted brains) that have all the phenomenological potential that characterizes the actual world, but in which there exist few, if any, of the physical things that actually exist; (2) the problem of *deceptive appearances*: idealism has no defensible way to distinguish between veridical appearances (like the ones we have in normal waking life) and deceptive appearances (like those that characterize dreams, hallucinations, and misperceptions); and, (3) the problem of *imperceptibles*: idealism is incompatible with the idea that spacetime contains imperceptible objects and events. In Chapter 9, I reply to these objections, arguing that none of them succeeds against a properly formulated idealism.

Chapter 10: The conscious self

The purpose of Chapters 10 and 11 is to deepen the phenomenalist position, by arguing that a phenomenalist need not regard either the conscious self (Chapter 10) or mental representation (Chapter 11) as an explanatory primitive.

In Chapter 10, we consider the main reductive analyses of the conscious self presently on offer (which seem to exhaust the theoretical possibilities in this area). It turns out that each of these can be formulated in a way that is compatible with radical phenomenalism. However, it also turns out that each of the analyses is a failure, at least when interpreted as an attempt to capture what we mean whenever we talk about conscious selves (subjects, minds).

From this, one might conclude that conscious selfhood is metaphysically or conceptually irreducible, or that conscious selves are somehow mere fictions. However, I argue that a more reasonable conclusion to draw is that we mean somewhat different things by "conscious self" (or "same conscious self") on relevantly different occasions. In this view, each of the failed analyses of the conscious self succeeds as an articulation of one of the several things we mean on different occasions when speaking of conscious selves. This particularist conception of conscious selfhood is consistent with phenomenalism, since phenomenalism is consistent with each of these several meanings.

Chapter 11: Intentionality

Radical phenomenalism analyzes physical things in terms of the interpretive possibilities of hypothetical totalities of conscious experiences. Interpretation, however, is an intentional concept: for a totality of experiences to be interpretable as including a perception of *x* is, at least in part, for it to be reasonable to *think* of that totality as including such a perception. And thinking belongs to the domain of mental representation (original intentionality).

Some idealists treat intentionality as an analytical primitive, alongside conscious experience itself. That is not the approach I take. Instead, I suggest that it is possible to reduce facts about mental representation to facts of a broadly phenomenological nature. The key intuition behind this suggestion is that it is impossible for two beings to duplicate one another in terms of their phenomenal dispositions—their dispositions to have one kind of conscious experience upon having another kind of conscious experience—without also duplicating one another in terms of the representational contents of their mental states.

A main source of resistance to this kind of account is the belief, dominant in recent philosophy, that the representational contents of a person's mental states often depend on how that person relates to his social and physical environment. In Chapter 11, I argue that this externalist view of mental representation is not supported by the arguments commonly advanced in favor of it, and that it is actually undermined by a careful consideration of the relationship between the intentional and phenomenological aspects of a person's mental life.

2

Space, time, and spacetime

> Supposing that last night I observed through my telescope a star that is, let us say, 100 light years away, not only did I know that the light entering my eye was emitted from that star 100 years ago, but also the star (or its individual atoms) knew 100 years ago that I, who did not even exist yet, would observe it last night at such-and-such a time, through a telescope having certain dimensions and lenses with certain refractive indices, looking in a certain direction, and using this eye.
>
> Hugo Tetrode, *On the Causal Interconnectedness of the World*

2.1 Introduction

Consider the two maps depicted in the figure below:

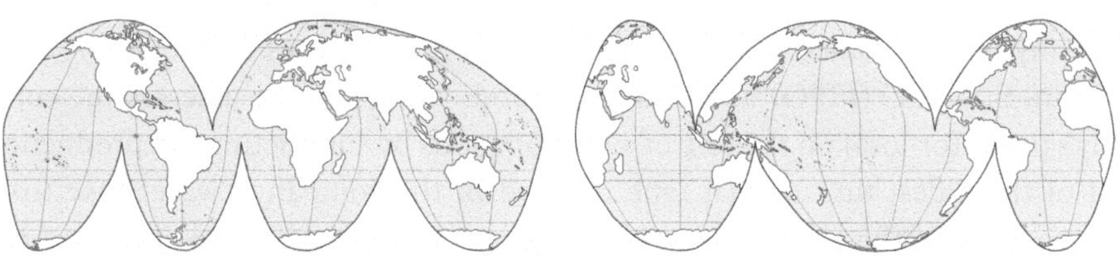

Continental Projection — Oceanic Projection

Figure 2.1 Information-equivalent projections of the Earth's surface.

These maps contain the same information—they are both equal-area projections exhibiting identical distortion—although some of the information is easier for us to extract from one map, some easier to extract from the other. A mariner plotting a course across the Pacific Ocean, for example, might prefer the oceanic projection, while a hiker planning a trek across North America might prefer the continental projection. But the oceanic projection's depiction of North America by means of two disconnected shapes does not correspond to any feature of North America—the oceanic projection is not *saying anything about* North America that the continental projection is not. The differences between the two projections are not differences in what the maps say about the planet they represent, but only in how they say it.

Assuming that these maps are accurate, every accurate map of the Earth's surface must have certain features in common with them. Given that the two maps are correct to depict Madagascar as separate from the African mainland, every other map of equal resolution must also depict Madagascar as separate from the mainland if it is to be counted as accurate.

Other features of the maps in Figure 2.1 need not be present in every accurate depiction of our planet. An equally accurate world map could use gray for land and white for water, or any other two-color scheme. The maps in Figure 2.1 have a bent or kinked appearance at 44°40′ North and South; this is an artifact of this particular projection (the Goode homolosine), and is absent in other, equally accurate cartographic representations of the Earth.

The two maps' depiction of Madagascar as separate from the African mainland answers to an absolute feature of the Earth: a feature that every other world map of comparable detail and accuracy exhibits. The maps' use of a particular color scheme does not answer to any terrestrial feature, but is merely an artifact of these particular maps' way of conveying the information about the Earth that they do.

The distinction between ontological absolutes and representational artifacts also arises in the business world. An accountant in charge of documenting a company's performance has various options in how he does so. He might use a "first-in, first-out" (FIFO) accounting method. Using this, he assesses the company's profit (or loss) on the sale of a unit of goods by subtracting the production cost of the oldest unit of that type of good in the company's warehouse from the proceeds of the sale. Alternatively, the accountant can use a "last-in, first-out" (LIFO) accounting method. Using this, he assesses the company's profit (or loss) on a unit sale by subtracting the production cost of the *newest* unit in the warehouse from the proceeds of the sale.

The accountant's choice of method is completely independent of the actual ages of items in the warehouse, and of the actual ages of the units sold to customers. He can use the LIFO method even if the company always sells units in the order in which it produces them, and he can use the FIFO method even if the company always sells its most recently produced units first. An accountant using the FIFO method will be in complete agreement with one using the LIFO method when it comes to the actual production dates of the items in the warehouse, the actual production costs of those items, the amount of money realized by each unit sale, and the amount of money in the company's bank account at any given time.

Where they differ is in how they describe the company's profits within a given period of time. In an inflationary setting, the FIFO accountant will record higher profits than the LIFO accountant from the sales completed in a given period. This is because the FIFO accountant assesses the production costs of the units sold as lower (since, by his method of accounting, these units were produced relatively farther in the past, when raw materials were cheaper). In a period of deflation, the FIFO accountant will record lower profits from sales completed in a given period, since by his method of reckoning, the units sold in this period were produced at a time of higher production costs.

Both accounting methods are accepted as consistent with good business practice, provided that they are applied consistently and uniformly across the business. Both give an equally clear picture of the state of the business. The fact that the FIFO method represents the business as having (say) higher profits in a given quarter than the LIFO method would have represented it as having is purely an artifact of representation. It is one way among other, equally acceptable ways of describing the same underlying commercial state of affairs.

Of course, there may be pragmatic reasons to prefer one method over the other. A publicly-traded company in an era of inflation may practice FIFO accounting in order to make its shares appear more valuable to unwary investors, or it might adopt the LIFO method in an effort to lighten its tax burden. But whichever method the company uses, its accountants' reports are equally informative and equally accurate descriptions of the actual, objective state of the company.

The distinction between ontological absolutes and representational artifacts also arises in natural science. From the earliest days of scientific inquiry, people have recognized that a scientific description of the world contains some elements that the world described demands to be included, and other elements that are in some sense arbitrary, conventional, or negotiable: features of the description demanded more by our particular needs and interests than by the world itself. A physical description's use of a particular system of measurement—feet versus meters, say—is an especially obvious example of an artifactual element. We shall encounter other, more interesting examples by and by.

Over the centuries, the scientific trend has been to move descriptive elements from the "absolute" column over to the "artifactual" column: to regard less and less of our descriptive activity as answering directly to What There Is, and more and more of it as a free choice among equally good routes to the bottom of things. The shift has been gradual, spread out over many generations, and has required tremendous intellectual effort. In the twentieth century, the trend led to the relativistic conception of time and space.

According to the relativistic conception, the world contains considerably less structure than meets the eye. In relativistic spacetime, the distance between spatially separated events, and the temporal order in which spatially separated events occur, are not, in general, absolute features of the world. For example, on the relativistic conception, the simultaneous occurrence of events at different spatial locations is not an absolute feature of reality, but an artifact of a particular representation of reality—one method of accounting for the underlying facts among many (in fact infinitely many) equally good methods.

The relativistic conception of spacetime has its roots in the Newtonian conception of time and space, which in turn is a natural development of our ordinary spatial and temporal concepts. But it would be a mistake to think that the Newtonian conception is nothing but a cleaned-up version of common sense. The Newtonian conception of time and space is firmly rooted in common sense, but the roots run deep, and their connection to the Newtonian system is not open to casual inspection. Indeed, most of the conceptual development in our understanding of time and space takes place in this first transition, from common sense to Newtonian spacetime. The later transition from Newtonian to relativistic spacetime involves a comparatively modest—although still very great—conceptual advance.

In this chapter, we trace back the Newtonian conception of time and space to its basis in everyday thought and practice, and then see how the relativistic conception of spacetime grows out of the Newtonian conception in response to certain empirical developments.

2.2 Distance, duration, and simultaneity

Our ordinary concept of simultaneity is bound up with what we ordinarily mean by stating that two or more events occur simultaneously, and nothing reveals what we mean by such statements better than the considerations that we offer and accept as evidence in support of them.[1]

Sometimes we judge that two events occur simultaneously on the grounds that they are two sides of an interaction. If two balls collide, we judge that the first struck the second at the same time as the second struck the first; otherwise, there wouldn't have been a collision. Similarly, if a book rests on a table, we judge that the book presses down on the table at the same time as the table supports the book. These are cases in which two events occur not only at the same time, but also at the same place.

[1] Here I am not pledging allegiance to verificationism; see Appendix.

More often we make judgments of *distal* simultaneity: judgments to the effect that two or more spatially separated events occur, or occurred, or are going to occur, at the same time. One possible basis for such a judgment is that we see (or hear or feel) the events at the same time. Lying behind this rationale is the assumption that the events are equally far away from us, and that news of them—carried by light, sound, or some other means—travels to us at the same speed. Because light in particular travels so fast, and because our field of vision is relatively limited, we normally don't bother to make this assumption explicit when we make a judgment of simultaneity based on sight. For everyday purposes, we tend to assume that light travels infinitely fast. When it comes to sound and other messengers, we are more circumspect. As children, we learn how to estimate the distance of an approaching thunderstorm by counting off the seconds between flashes of lightning and subsequent peals of thunder, and with this comes the recognition that sound from an earlier event (the thunderbolt) may reach us later than sound from a later event (our counting out loud).

What about events that occur outside of our immediate perceptual environment? Our reasons for thinking that two such events take place at the same time often have to do with how the events relate to clocks and calendars.

Our earliest clocks were the Sun, Moon, and stars. These, however, were of limited use in establishing distal simultaneity. The angle between a celestial body and the horizon is approximately the same only over limited stretches of the Earth's surface: high noon in London is not the same time as high noon in Moscow.

More accurate and reliable judgments of simultaneity had to wait for the development of portable clocks that could be synchronized and depended upon to stay that way in transit. In the late eighteenth century, such clocks finally became available, in the form of accurate marine timepieces. With these, it was possible to synchronize clocks by direct optical or mechanical means in the same workshop, and then move them to new locations without interfering with their operations. Unlike pendulum clocks, you could transport one spring-driven marine chronometer from Greenwich to Barbados, and another from Greenwich to Cairo, without throwing them off kilter. In combination with identical calendars, people could then judge that an event *B* in Barbados occurred at the same time as an event *C* in Cairo, on the grounds that the clock in Barbados displayed the same time on the same day when *B* took place as the clock in Cairo displayed on that day when *C* took place there.[2]

Another way to synchronize clocks became possible with the advent of telegraphy and other forms of high-speed, long-distance communication.

[2] Gould (1923) provides a detailed account of the evolution of marine chronometers.

Nineteenth-century telegraph operators working at far-flung locations synchronized their clocks using a now familiar method (which is the basis of the Network Time Protocol that your computer uses to set its clock to server time). Andre, in Lima, let's say, sends a message to Marie in Paris, who has a clock of the same construction as Andre's. Andre's message asks Marie to reply to Andre in exactly 10 seconds (as measured by Marie's clock) after Marie receives Andre's message. Suppose Andre's clock reads 13:00:00 when he sends his message to Marie. When Marie receives the message, her clock (in Paris) reads, let's say, 19:07:35. When her clock reads 19:07:45, she sends her reply back to Andre in Lima, informing Andre that she sent her message ten seconds (by her clock) after receiving his. Suppose that when Andre receives Marie's reply, his clock reads 13:00:12. Andre surmises that it took 1 second for his telegram to reach Marie in Paris, and 1 second for Marie's message to travel from Paris back to Lima. Andre now waits until his clock reads 13:01:12, whereupon he sends Marie another message, this one instructing her to wait one minute (as measured by her clock), and then set her clock to 13:02:13. Marie follows Andre's instructions, and the operators now judge their clocks to be synchronized.[3]

In adopting this method of synchronizing their clocks, the telegraphers assume that electrical signals travel along the cable between them at the same speed in either direction, and regardless of when the signals get sent. In technical terms, this is the assumption that telegraphic transmission is "isotropic" and "homogeneous"—isotropic, in the sense that the speed of the signal is independent of the direction in which it travels (e.g. East versus West), and homogeneous in the sense that the speed of the signal is independent of when and where the signal originates (e.g. in Lima versus Paris, or at local noon versus local sunset).

Assumptions of signal isotropy and homogeny underlie many systems of measurement. Early, and remarkably accurate, measurements of the Earth's size were expressed in terms of "stadia," a "stadion" being one one hundredth of the distance traversed by a camel caravan in one day. These measurements were based on the assumption that a camel caravan covers the same distance from one sunrise to the next, regardless of the direction in which it travels, and regardless of where it begins the day's journey. In theory, one could use camels to synchronize clocks, although this would require averaging over many round trips, and could never achieve more than an approximate synchrony, given the vicissitudes of desert travel.

[3] The telegraphic synchronization of clocks was an historically important example of so-called Einstein synchronization; for details, see Galison (2003), 174–91.

Especially crucial to modern descriptions of the world in spatial and temporal terms is the principle that *inertia*—resistance to change in state of motion—is homogeneous and isotropic. It is hardly an exaggeration to say that this principle is the foundation of all modern physics. Newton adopts the principle explicitly with his first two laws of motion. The first law states that an unimpressed object—one free from influence by external forces—does not change its state of motion for as long as it remains unimpressed: its state of motion is independent of its position in space and time. This comes to the assumption that an object travels at constant speed along a straight line (or else does not move at all) as long as it is free from influence by external forces. The second law states that the force required to change the speed of an object having a given mass by a given amount is independent of the direction in which the force gets applied, and independent of when or where in the world the application of force takes place.[4]

We do not arrive at the principles of inertial homogeny and isotropy by studying the behavior of unimpressed objects, since we never observe unimpressed objects: the idea of an object that is uninfluenced by *any* external force is an idealization. But we can discover how unimpressed objects *would* behave, by studying the behavior of objects under progressively less impressed conditions. If we notice a pattern in this behavior that tends toward some limit, we may infer that that is how a truly unimpressed object would behave, if there were such an object.

For example, Galileo arrived at the conclusion that an unimpressed object moves (if at all) at a constant speed, by noticing that the more he isolated a rolling ball from external forces, the more the ball tended to move at a constant speed. Reflections on the behavior of bodies moving around in the hold of a ship led him to surmise that it takes equal force to move equally massive objects equal distances in equal times, regardless of when, where, or in what direction the force gets applied.

In arriving at his conclusions about inertial behavior, Galileo obviously had to use some ready-to-hand way of describing distances and times in his experiments, such as yardsticks and plumb lines (for distance), and water-clocks or his own pulse (for time). But his conclusions inspired another method of describing phenomena in spatial and temporal terms, which would supersede the more homely methods available to him. This is the method of describing phenomena in terms of an inertial coordinate system.[5]

[4] See Newton (1686/1934), 13–14. The purpose of Newton's third law of motion is to define the concept of equal (or balanced) force that features in the first two laws.

[5] For Galileo's methods, see Galileo (1638/1954), 179.

2.3 Inertial coordinate systems

The basic idea behind an inertial coordinate system is to assign coordinates to events in such a way that each unimpressed object gets construed as moving at a constant speed in a straight line (or else as standing still), and also in such a way that for any given mass (quantity of matter), the force required to accelerate an object having that mass to a certain speed is the same, regardless of the direction of acceleration, and regardless of where and when the acceleration takes place. In short, an inertial frame of reference is a way of assigning spatial and temporal coordinates to events so that when we describe the world in terms of these coordinates, we construe inertia as homogenous and isotropic.

It turns out that when we describe the world in terms of an inertial frame of reference, we describe it as a world in which observed phenomena conform to a small number of simple laws of change, such as Newton's law of gravitation and Maxwell's laws of electromagnetism. Described in terms of an inertial coordinate system, all things move and change with a wonderful regularity. The existence of coordinate systems in terms of which our world has this regularity—or any discernible regularity—is not to be taken for granted. It is remarkable that such coordinate systems exist. But they do, in the form of inertial coordinate systems.

In practice, identifying and using an inertial frame of reference can be a laborious undertaking. Early descriptions of the motions of the planets in terms of inertial coordinates required collating and analyzing decades' worth of astronomical data. To fix ideas, it will be helpful to illustrate the concept of an inertial frame with a simpler, fictional example.[6]

Suppose we have eight space stations and a mother ship—call her *Nike*. *Nike* (as well as the eight stations) bristles with guns pointing out into space in all directions, which, when fired, eject equally massive projectiles with equal force. All the stations also have identically constructed onboard clocks. Every station is manned by a person who is instructed to fire all his station's guns whenever a projectile from *Nike* passes by his station.

We place the eight stations around *Nike*, and reposition them until the following state of affairs obtains: whenever *Nike* fires all her guns at once, she later receives incoming projectiles from all the outlying stations simultaneously, where the time (as measured by *Nike*'s clock) between *Nike*'s successive outgoing salvoes and successive incoming reply salvoes is always the same.

Intuitively, this means that the eight surrounding stations are at a constant equal distance from *Nike*. What backs up this intuitive judgment is our assumption—

[6] The thought-experiment that follows is inspired by the set-up described in Brown (2010), 25.

inspired by observations such as Galileo's—that all the projectiles travel at the same, constant speed in straight lines, regardless of the fact that they are traveling in different directions, and being fired from different locations and at different times.

Next, we set all the stations' clocks using much the same method as Marie and Andre used to synchronize their clocks via telegram. *Nike* fires a salvo in all directions when her onboard clock reads t_1—to make it more concrete, let's say that t_1 is 13:00:00. When she receives ballistic replies from the surrounding stations, her clock reads, let us say, 15:00:00. She now loads her guns with projectiles just like the usual ones, except that they have the following message etched on them: "Set your clock to 17:00:00." When her onboard clock reads 16:00:00, *Nike* fires the etched projectiles. When the personnel aboard the surrounding stations see these projectiles pass by, they read the message etched on them, and set their clocks accordingly.

Intuitively, this method of setting the stations' clocks results in the clocks being synchronized. What supports this intuition is, again, our assumption that equally forcible pushes accelerate equally massive objects to equal velocities regardless of the direction of the pushing, and that the trajectory of an unimpressed object is straight and speed-wise constant.

Next, we arrange the eight outlying stations so that each of them is equidistant from exactly three other outlying stations, equidistance being established by the same method we used to position the outlying stations at an equal distance from *Nike*. Intuitively, this means that the eight outlying stations form a cube, with one station located at each corner, where this cube is inscribed in a sphere having *Nike* at its center (see Fig. 2.2). We label one of the outlying stations N(0,0,0), and the three stations from which it is equidistant N(−1,0,0), N(0,−1,0), and N(0,0,−1), respectively.

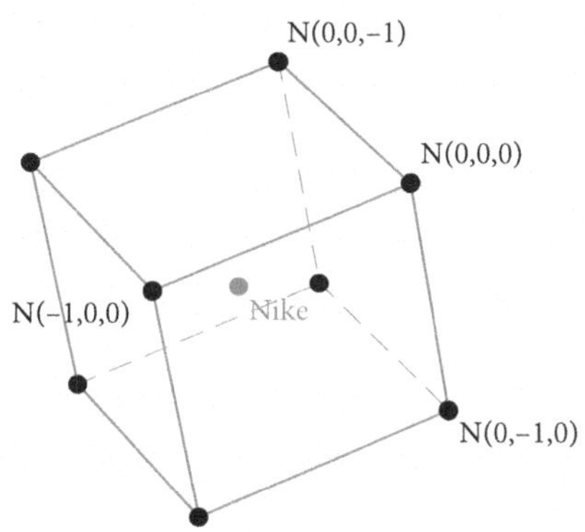

Figure 2.2 *Nike* surrounded by eight stations.

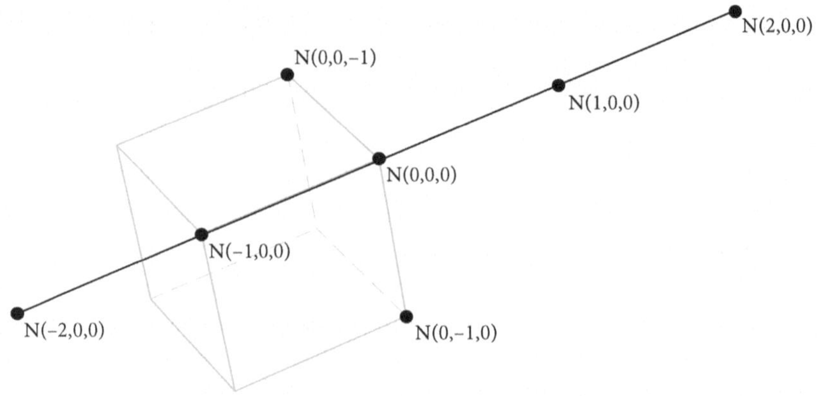

Figure 2.3 Constructing the *x*-axis.

We now introduce three new stations to the array: N(−2,0,0), N(1,0,0), and N(2,0,0); see Figure 2.3. We maneuver these stations around until the following four conditions are satisfied: (i) any salvo fired from N(2,0,0) that passes right by N(1,0,0) also passes right by N(0,0,0), N(−1,0,0), and N(−2,0,0); (ii) N(0,0,0) receives return salvoes from N(−1,0,0) and N(1,0,0) simultaneously; (iii) N(−1,0,0) receives return salvoes from N(−2,0,0) and N(0,0,0) simultaneously; and (iv) N(1,0,0) receives return salvoes from N(0,0,0) and N(2,0,0) simultaneously.

We synchronize the clocks of the newly-introduced satellites with the onboard clock of station N(0,0,0) using the same method as before. Intuitively, we have now established an evenly-graduated and synchronized x-axis. To establish y- and z-axes, we apply the same procedure in relation to N(0,−1,0) and N(0,0,−1); see Figure 2.4.

These diagrams, while hopefully useful, are also potentially misleading, since they can give the false impression that we are arranging the stations until they look like a rectangular grid of satellites with clocks ticking in sync. This could give the further false impression that we can check the array by visual inspection to make sure the axes are mutually perpendicular, the stations evenly spaced, and the clocks synchronized. But any such "checking" would make implicit assumptions about how long it took optical signals from the stations to reach the "checker," which would, in turn, bring in assumptions of signal isotropy and homogeny. The only way to break away from such assumptions would be to get a view of the array mediated by infinitely fast signals, but such signals do not exist in our universe, or, if they do, they are not available for us to use. So we should not let the diagrams obscure the fact that in setting up the array, we define co-linearity, perpendicularity, equidistance, and simultaneity in purely inertial terms.

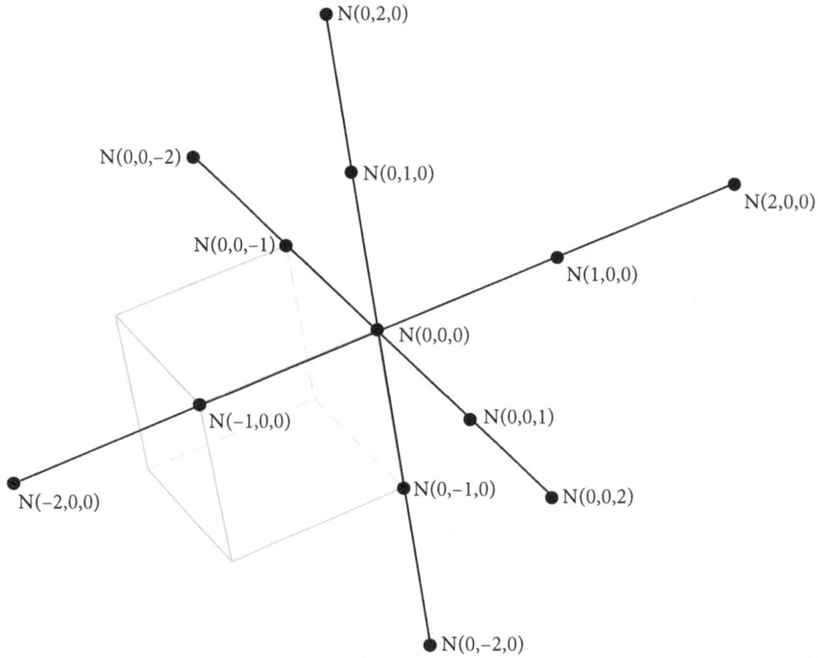

Figure 2.4 Adding *y*- and *z*-axes.

By repeating the above-described method of positioning stations as many times as we like, we can establish a rectangular array as extensive as we like; and by positioning the stations close enough together, and using clocks with high enough rates, we can graduate the spacetime axes as finely as we like.

With such a system of stations in place, we can now assign coordinates to events as follows: for any event *E*, and any station S in the *Nike* array, if *E* occurs at S when S's clock reads *t*, *E*'s coordinates are (x,y,z,t), where *x*, *y*, and *z* are the numbers occurring in S's name. For instance, if a piece of interstellar debris glances off of station N$(a,-b,c)$ when that station's onboard clock reads *d*, this event (of the debris striking the station) gets assigned coordinates $(a,-b,c,d)$.

This method of assigning coordinates to events is more or less an inertial coordinate system—more or less, but not strictly speaking. Strictly speaking, the inertial coordinate system in this case consists of the described Nikean coordination of events *plus* every other possible coordination in terms of which the satellites in the *Nike* array are persistently at rest. One example is a rectangular coordination that places a different satellite in the *Nike* array at its geometric origin. Another is a polar coordination based on an array of inertially synchronized satellites extending outward from a given satellite in an inertially-established starburst pattern, where these satellites are at rest with respect to the satellites in the *Nike* array. And there are other possibilities as well.[7]

In Newton's world-view, simultaneous events are just events that take place (or could have taken place) in the presence of inertially-synchronized clocks displaying the same reading. A process that takes ten seconds to complete is just a process whose first event takes place (or could have) in the presence of a clock reading t, and whose last event takes place (or could have) in the presence of an inertially-synchronized clock reading t-plus-ten-seconds. A stick ten units long is a stick that we can position in such a way that one of its ends lies at a space station whose clock reads t, and its other end lies at a space station inertially positioned ten space stations away, whose inertially-synchronized clock also reads t. And so forth.

2.4 Relativistic spacetime

The idea of using the principle of inertia to map space and time marked a revolution in our understanding of the physical world. But it left much unchanged. Concerning a couple of very basic points, Aristotle and Newton are in perfect agreement. Both believe that the distance between two objects at a given time is an absolute fact about those objects, and both believe that the time interval between any two events is an absolute fact about those events. According to the Newtonian conception of space and time, if any complete and accurate description of the world entails the existence of a complex of events occurring simultaneously at various distances from one another, then every complete and accurate description of the world entails the existence of that very same complex of events. The modern metaphysician's "time-slices" of reality have the status of absolute existents, on the Newtonian view. Unfortunately for those with a fondness for time-slices, this is a point on which the Newtonian view proves to be mistaken.[8]

[7] Many authors speak of "observing an event from a given frame of reference." Such talk is extremely misleading: it makes no sense to speak of observing something from a frame of reference. A frame of reference is not a vantage point; it is simply a way of assigning coordinates to events, and so a way of *describing* events in mathematical terms. An "inertial" frame, as we have seen, is a way of assigning coordinates to events in conformity with the principle that inertia is homogenous and isotropic. It is no more possible to observe things from an inertial frame of reference than to observe things from a map of France. Unless you are a Lilliputian, or own a really big atlas, you cannot observe anything from a map of France.

Many authors also speak of "an observer's frame of reference." This is also misleading. There is no such thing as "the frame of reference of a given observer," since an observer is generally describable in terms of any one out of infinitely many frames of reference.

[8] My discussion in this section is indebted to Einstein (1905), Minkowski (1909/1952), Pauli (1958), Mermin (1968), Geroch (1978), and the bibliographically obscure but admirably lucid Brown (2010).

We describe phenomena as conforming to the same laws, provided that we describe them in terms of some inertial coordinate system; *which* inertial coordinate system we use doesn't matter. This is the content of the Galilean principle of relativity—"Galilean" because it was Galileo who first articulated it, and (less happily) "relativity" since the principle states, in effect, that you describe the world as an equally orderly and lawlike place, regardless of whether you describe it "relative to" (i.e. in terms of) one versus another inertial frame of reference.

The principle of relativity says that inertial frames of reference yield equally illuminating descriptions of Nature, but beyond that, it implies nothing about how different inertial frames of reference compare to one another. One respect in which two inertial frames can *differ* from one another is by "moving" with respect to one another. If an unimpressed object is at rest as described in terms of one inertial frame, it moves at a constant speed in a straight line as described in terms of a different inertial frame. More generally, for any object at any time, there is some inertial frame in terms of which that object is at rest at that time, and some inertial frame—in fact, infinitely many—in terms of which it is in motion at that time.

Different inertial frames of reference yield descriptions of reality that do an equally good job of making the motions of bodies intelligible in terms of simple laws, despite the fact that one of the descriptions describes a given object as stationary at a given time, while others describe it as moving at that time. We can put this by saying that there is no "fact of the matter" about whether a thing is moving or at rest at a given time. To the extent that we can speak of an object's momentary state of motion at all, its state of motion (or rest) at a given moment is a mere artifact of representation. It is a representational artifact, in the sense that if there is a complete and accurate description of reality in terms of which the object is at rest at a given time, there is an equally complete and accurate description of reality in terms of which the object is then in motion.[9]

The claim that different inertial coordinate systems afford equally complete and accurate descriptions of the world is not entirely free from controversy. There are, for example, some who say that an object has a "true" state of motion given by its change of position relative to an unobserved "absolute space." If that is correct, then only one inertial coordinate system permits a complete and accurate description of reality, since only one system is such that objects at rest relative to

Throughout, the focus is on special relativity. This theory is false, but the features of it that are relevant to my purposes survive into the theory of general relativity which supersedes it.

[9] By a "complete and accurate" description of reality, I mean a description that tells the truth, the whole truth, and nothing but the truth (at least as regards concrete, non-mathematical reality). There are various ways of making this idea more precise—for extended discussion of the relevant issues, see Chalmers (2012).

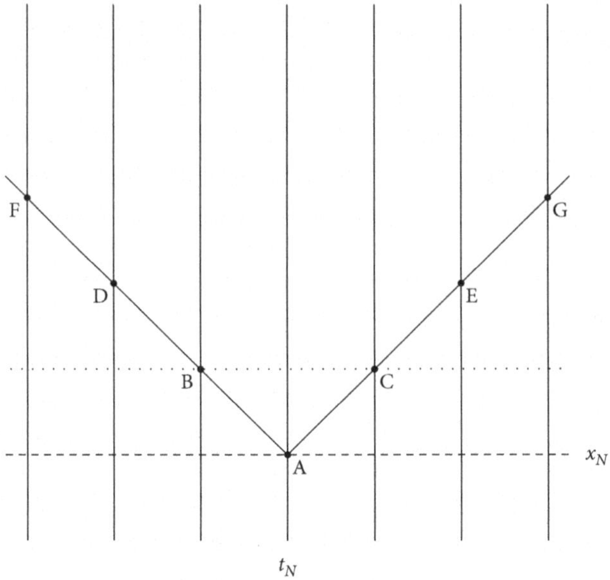

Figure 2.5 Light propagating through *Nike* array, in terms of Nikean coordinates.

absolute space are at rest in terms of that system. I take up this issue in an Appendix at the end of the book. Here, I adhere to the usual view that a complete and accurate description of the world is possible in terms of any inertial frame of reference.

Two inertial frames of reference may differ from one another by describing the same object as moving or at rest, as well as in other respects. But there is one surprising respect in which all inertial frames of reference are alike: a photon or light wave has the same velocity in terms of every inertial frame of reference.

Suppose that station N(0,0,0) of the *Nike* array emits a flash of light. The other stations in the array are equipped with mirrors to reflect back any light emanating from station N(0,0,0). In terms of the Nikean frame of reference, the light propagates away from N(0,0,0) at the same speed in all directions: in terms of the Nikean frame, light from the flash reaches station N(−1,0,0) at the same time as it reaches N(1,0,0); reaches N(−2,0,0) at the same time it reaches N(2,0,0); etc. Experimentally, this is borne out by the fact that station N(0,0,0) receives light reflected back from N(−1,0,0) at the same time it receives light reflected back from N(1,0,0); from N(−2,0,0) at the same time as from N(2,0,0); etc., as well as by the fact that there is a clock-reading, $t+1$, such that both N(−1,0,0) and N(1,0,0) report that their clocks read $t+1$ when the light emitted by station N(0,0,0) reached them; a clock-reading $t+2$ such that both N(−2,0,0) and N(2,0,0) report that their clocks read $t + 2$ when the light emitted by N(0,0,0) reached them; etc.

In short, in terms of Nikean coordinates, station N(0,0,0) lies persistently at the center of an expanding spherical shell of light. We can depict this situation by means of a spacetime diagram (see Fig. 2.5). In this diagram, vertical lines depict things as remaining stationary over time, and lines that cross the dashed horizontal line (labeled x_N) at a 45 degree angle represent things that move at *c*: the speed of light *in vacuo*. The vertical line that passes through A depicts the Nikean station N(0,0,0) as remaining at rest over time; the vertical line passing through B depicts N(−1,0,0) as remaining at rest; the vertical line passing through C depicts N(1,0,0) as remaining at rest; etc. Only stations along the *x*-axis of the *Nike* array [the axis extending from N(−2,0,0) to N(2,0,0) in Fig. 2.4] are depicted here. A line crossing the x_N-axis at an angle of less than 90 degrees but greater than 45 degrees would represent the trajectory of an object traveling at less than the speed of light along the *x*-axis of the *Nike* array (in one direction or the other).[10]

Labeled dots in the diagram depict events of interest to us. The diagram depicts two events as occurring simultaneously by placing their respective dots in the same line parallel to the x_N-axis. In this diagram, therefore, events B and C are represented as simultaneous, as are events D and E, and events F and G (to make things vivid, I have included a dotted line to emphasize that the points labeled "B" and "C" are equidistant from the x_N-axis). Figure 2.5 depicts the following events:

A: Station N(0,0,0) emits flash of light; onboard clock of N(0,0,0) reads t.
B: Light from flash reaches station N(−1,0,0); onboard clock of N(−1,0,0) reads $t + 1$.
C: Light from flash reaches station N(1,0,0); onboard clock of N(1,0,0) reads $t + 1$.
D: Light from flash reaches station N(−2,0,0); onboard clock of N(−2,0,0) reads $t + 2$.
E: Light from flash reaches station N(2,0,0); onboard clock of N(2,0,0) reads $t + 2$.
F: Light from flash reaches station N(−3,0,0); onboard clock of N(−3,0,0) reads $t + 3$.
G: Light from flash reaches station N(3,0,0); onboard clock of N(3,0,0) reads $t + 3$.

[10] In the standard terminology, events that lie on the trajectory (actual or hypothetical) of a signal traveling less than the speed of light are "timelike separated," events that lie on the actual or hypothetical trajectory of a signal traveling at the speed of light are "lightlike separated," and events that do not lie on the actual or hypothetical trajectory of any luminal or subluminal signal are "spacelike separated."

We are justified in placing B and C on the same horizontal line by the fact that the onboard clocks of N(−1,0,0) and N(1,0,0) display the same reading when the light from N(0,0,0) reaches them, these clocks having been synchronized as described earlier. (We have a similar justification for placing D and E, and F and G, on the same horizontal lines.)

To represent the fact that station N(0,0,0) receives reflected light signals from N(−1,0,0) and N(1,0,0) simultaneously [and likewise for N(−2,00) and N(2,0,0), N(−3,0,0,) and N(3,0,0), etc.], we can add further events to the diagram. In Figure 2.6, H is the simultaneous receipt by N(0,0,0) of light signals reflected back from N(−1,0,0) and N(1,0,0), and I is the simultaneous receipt by N(0,0,0) of light reflected back from N(−2,0,0) and N(2,0,0). Diagonal lines represent the paths of various light signals from station to station.

Now suppose that we set up an array of space stations just like the *Nike* array (same guns, clocks, mirrors, etc.), except that we construct the array around a different mothership, *Perseus*. We position the stations in the *Perseus* array just as we positioned the stations of the *Nike* array, and synchronize the onboard clocks of the *Perseus* stations using the same technique we used to synchronize the Nikean clocks.

Perseus, like *Nike*, is free from influence by external forces, but the two ships are not stationary with respect to one another. Rather, in terms of the Nikean

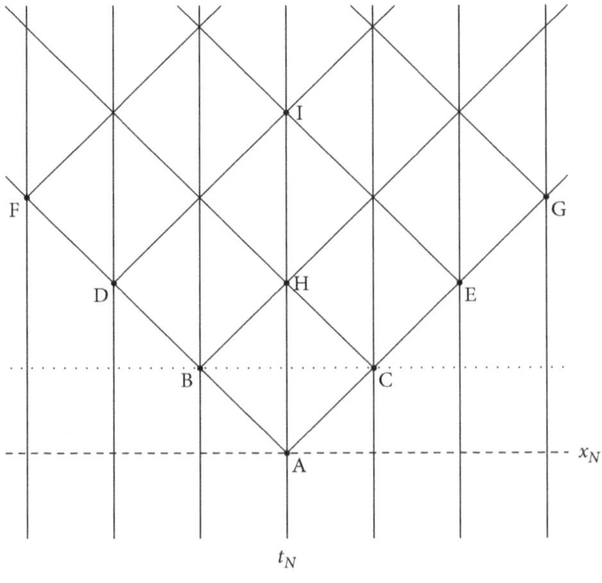

Figure 2.6 Light signals in the *Nike* array, in terms of Nikean coordinates.

coordinate system, *Perseus* and the stations in her array move at constant speed in the N(0,0,0) to N(1,0,0) direction.

Suppose that just as *Nike*'s station N(0,0,0) emits its flash of light, *Perseus*'s station P(0,0,0) crosses paths with N(0,0,0).[11] Suppose furthermore that the clock aboard N(0,0,0) shows the same reading, t, as the clock onboard P(0,0,0) when the two stations meet. In terms of Nikean coordinates, the situation is then as depicted in Figure 2.7, where oblique lines depict the trajectories of the stations that constitute the x-axis of the Perseid array, the lines crossing x_N at 45 degree angles represent the path of light emitted by N(0,0,0) when this station crosses paths with P(0,0,0), and the vertical line that passes through the dot marked "a" represents the location over time of N(0,0,0).

Empirically, we find that P(0,0,0) receives a light signal reflected back from P(−1,0,0) at the same time as it receives a signal reflected back from P(1,0,0), and likewise for P(−2,0,0) and P(2,0,0), etc.[12] To represent this fact, we add some

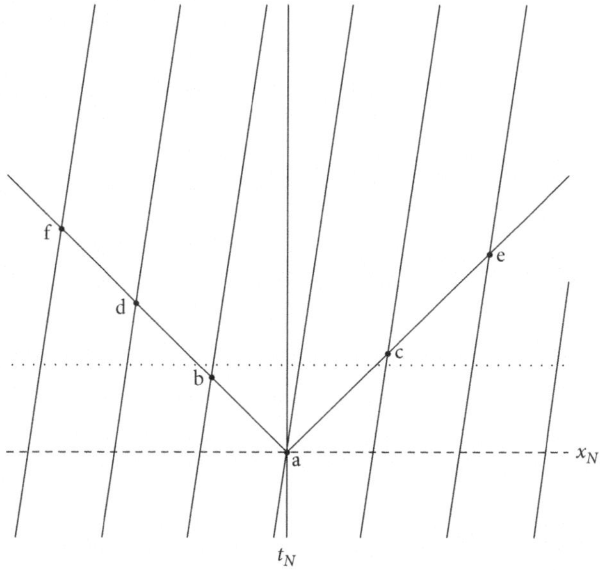

Figure 2.7 The *Perseus* array, in terms of Nikean coordinates.

[11] Naturally, we cannot have *Perseus* literally intersect with *Nike*—there would be a catastrophic collision if we tried to achieve that. But we can do multiple test runs or "fly-bys," involving successively closer station encounters each run, and extrapolate from the pattern that emerges what would happen if we could somehow make the *Perseus* and *Nike* arrays momentarily coincide without interacting.

[12] Of course, since the *Nike* and *Perseus* arrays are fictional, there are no actual empirical findings to speak of here. But the hypothetical findings have ample precedent in actual experiments, the most famous of which is described in Michelson et al. (1887).

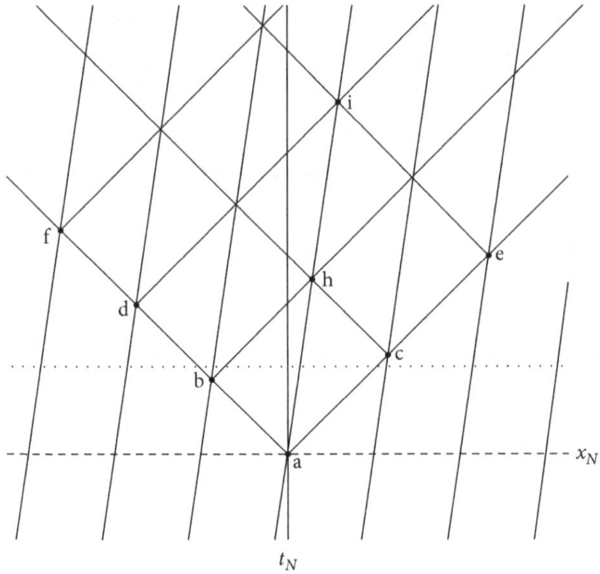

Figure 2.8 Light signals in the *Perseus* array, in terms of Nikean coordinates.

more 45 degree lines to the diagram, extending upward from b, c, d, e, etc. as in Figure 2.8; here is a list of the events depicted in that diagram:

- a: Station N(0,0,0) passes station P(0,0,0); N(0,0,0) emits flash of light; onboard clock of N(0,0,0) reads t; onboard clock of P(0,0,0) reads t.
- b: Light from flash reaches station P(−1,0,0); onboard clock of P(−1,0,0) reads $t + 1$.
- c: Light from flash reaches station P(1,0,0); onboard clock of P(1,0,0) reads $t+1$.
- d: Light from flash reaches station P(−2,0,0); onboard clock of P(−2,0,0) reads $t + 2$.
- e: Light from flash reaches station P(2,0,0); onboard clock of P(2,0,0) reads $t+2$.
- f: Light from flash reaches station P(−3,0,0); onboard clock of P(−3,0,0) reads $t + 3$.
- h: P(0,0,0) receives light reflected back from stations P(−1,0,0) and P(1,0,0).
- i: P(0,0,0) receives light reflected back from stations P(-2,0,0) and P(2,0,0).

The thing to take note of here is that P(−1,0,0)'s clock gives the same reading when light reaches its station as P(1,0,0)'s clock gives when light reaches *its* station. This, despite the fact that, in terms of the Nikean coordinate system, these events—that of light reaching P(−1,0,0), and that of light reaching P(1,0,0)—do not take place at the same time. In the diagram, this is shown by the fact that b occurs lower down than c: there is no line parallel to x_N that contains both b and c. The implication is that in terms of the Nikean frame of reference, the clocks of the Perseid array are out of sync.

Does this mean that there is something wrong with the clocks of the Perseid array? No: the clocks work just like those on the Nikean stations, and have been synchronized using the same method. Furthermore, there is complete agreement between those who assign coordinates to events using the Perseid system and those who assign coordinates to events using the Nikean system, when it comes to which events take place. In terms of either system, events A, B, C, D, E, F, G, H, I, a, b, c, d, e, f, h, and i take place, just as characterized above.

What, in terms of the Perseid frame of reference, is going on within the *Nike* array? In Perseid terms, *Nike* and her satellites move at constant velocity in the P(0,0,0) to P(−1,0,0) direction. In terms of the Perseid frame of reference, we can depict the events that occur in the *Nike* array as in Figure 2.9. Here, the vertical line passing through A represents the position of the Perseid station P(0,0,0) over time, and the oblique line passing through A depicts the position over time of the Nikean station N(0,0,0). The other oblique lines that cross the x_P-axis depict the positions over time of the other stations along the x-axis of the *Nike* array. The events represented in the diagram are identical to those depicted in Figure 2.6. In particular, "B" stands for the event of light reaching station N(−1,0,0) when that station's clock reads $t + 1$, and "C" for the event of light reaching station

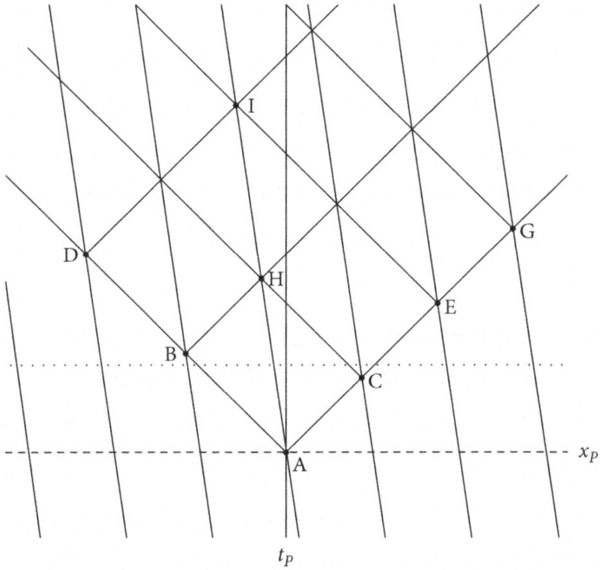

Figure 2.9 Light signals in the *Nike* array, in terms of Perseid coordinates.

N(1,0,0) when *that* station's clock reads $t + 1$. Since the diagram does not depict these events on a single line parallel to x_P, it depicts them as successive rather than simultaneous: first C, then B.

In terms of an inertial frame of reference that describes *Nike* as stationary, the light that N(0,0,0) emits propagates away from its point of origin with equal speed in all directions. In terms of an inertial frame of reference that describes *Perseus* as stationary, the same light propagates away from its point of origin with equal speed in all directions. But in terms of the Nikean frame of reference, what rests at the light's point of origin is station N(0,0,0), whereas in terms of the Perseid frame of reference, it is station P(0,0,0) that rests at the light's point of origin. All of this, despite the fact that station N(0,0,0) and station P(0,0,0) are not at rest with respect to one another (in terms of any inertial coordinate system).

What does this tell us about how these two inertial coordinate systems relate to one another? Among other things, it tells us that spatially separated events that are simultaneous in terms of one of the systems are *not* simultaneous in terms of the other. We describe the world as containing the same events conforming to the same laws, regardless of whether we describe them in terms of one versus the other coordinate system. Yet, when we describe phenomena in terms of one system, we describe a certain pair of events as simultaneous, whereas we describe the same events as successive when we describe phenomena in terms of the other coordinate system.

The same point applies to the durations of states or processes. Consider two sequences of events, such as the sequence involved in the journey of a photon from N(0,0,0) to N(1,0,0), and the sequence involved in the motion of a distinct photon from N(0,0,0) to N(−1,0,0). In terms of the Nikean coordinate system, these two journeys take the same amount of time: they have equal duration. (In Fig. 2.5, this is indicated by the fact that the dots representing events B and C are at an equal distance from the x-axis.) In terms of the Perseid coordinates, the journey from N(0,0,0) to N(−1,0,0) takes longer than the journey from N(0,0,0) to N(1,0,0) (as indicated by the fact that in Fig. 2.9, the mark for event B is farther above the x-axis than the mark for event C).

If we describe the world in terms of Nikean coordinates, we describe the journey from N(0,0,0) to N(1,0,0) as taking a certain amount of time. If we describe the world in terms of Perseid coordinates, we describe the same journey as taking a lesser amount of time. And, in general, for any sequence of events, it is possible to describe that sequence (in terms of one system of coordinates) as having a certain duration, and possible to describe the same sequence (in terms of a different system of coordinates) as having a different duration. Our choice to describe a given process as having one versus another duration is not fully constrained by the process itself, or by anything in the world we aim to describe. Choosing to describe the process as having one versus another duration is like choosing the continental projection in Figure 2.1 over the oceanic projection.

The situation here is exactly parallel to the case of absolute motion and rest. As we saw earlier, for any object at any time, there is some inertial coordinate system in terms of which that object is momentarily at rest at that time, and infinitely many other inertial coordinate systems in terms of which it is in motion at that time. Since we give an equally complete and accurate description of observed phenomena in terms of any of these reference frames, it is clear that only those features that all the reference frames have in common correspond to an objective feature of the world that we use the frames to describe. Any other feature that an inertial frame of reference may have, such as the feature of construing a given unimpressed object as moving, or the feature of construing a certain pair of spatially separated events as simultaneous, is simply a reflection of that frame's particular way of assigning coordinates to events so as to yield an intelligible picture of reality.

The choice to represent a pair of spatially separated events by means of time-identical versus time-distinct coordinates is analogous to a decision to represent the surface of the Earth by means of a continental versus oceanic projection, or the decision to account for a company's inventory using the FIFO versus the LIFO methods described at the beginning of the chapter. The FIFO accountant assigns the company one level of profits for a given financial quarter, the LIFO accountant another, but this doesn't change the fact that their financial reports record the same underlying information about the business. Likewise, descriptions of the world in terms of different inertial frames of reference may contain equal information, even though one of them assigns the same time-coordinate to each of two events, and the other assigns those events different time-coordinates. The difference between Figure 2.6 and Figure 2.9 is analogous to the difference between the two maps in Figure 2.1. In both cases, the differences are not differences in what the representations say about the world they represent, but only in how they say it.

Of course, not *everything* is a mere artifact of representation, on the relativistic view. For example, if any complete and accurate description of the world describes it as a place in which there is a causal connection between two given events, then *every* complete and accurate description of the world must do so. Likewise, if any complete and accurate description the world portrays two objects as colliding, every complete and accurate description of the world must do so. Collisions and causal chains of events are *not* mere artifacts of particular ways of representing the world: they are entirely real parts of the world we represent.

More generally, it is an absolute feature of the world that its constituent events exist at certain distances from one another in *spacetime*. Separate events occur at the same spacetime distance from one another in terms of every complete and accurate description of the world. This, despite the fact that the events have no absolute spatial distance from one another, and no absolute temporal distance from one another.

2.5 Conclusion

Henri Poincaré concludes his famous essay on the measure of time with the following words:

> The simultaneity of two events, the order of their succession, or the equality of two durations, should be defined such that the statement of natural laws is as simple as possible. In other words, all these rules, all these definitions are nothing but the fruit of an unconscious opportunism.[13]

The barbed coda is unfortunate. Poincaré is right to insist that Nature does not force us to adopt any particular conception of simultaneity or duration—or spatial co-location or distance, for that matter. By "simultaneous," we choose to mean something defined in terms of inertially-synchronized clocks because we understand what it is for clocks to be inertially-synchronized, and we find that when we describe events by reference to inertially-synchronized clocks, what emerges is a picture of the universe that is simple and intelligible to us.

In this sense, duration and simultaneity are "conventional": we choose to employ the temporal concepts we do because employing those concepts yields results that are satisfying to us.

But while our decision to define spatial and temporal concepts in the way we do is justified only by the results we get from making that choice, it is *not* up to us that a choice that yields such results is possible. There is nothing conventional about the fact that there exists a principle, such as the principle of inertia, that we can both grasp and use to construct a map of the world that is readily intelligible to us. The world did not have to be this way, and if it hadn't been, no amount of "unconscious opportunism" could have made it this way. The fact that we can make sense of the universe in terms of laws that are simple and intelligible to us is not one for which we can take credit. It is, as it were, a gift to us from the universe.

A contemporary theologian ignores Darwin's theory of evolution at his own peril. Even if he suspects that evolution has no bearing on his religious views,

[13] Poincaré (1913), 234.

he would be complacent not to assure himself of this before attempting any serious foray into natural theology. A contemporary metaphysician who assumes a Victorian conception of time and space is guilty of a similar complacency. Even if post-Newtonian developments in our understanding of time and space should prove to have no ultimate bearing on the nature of conscious experience and its relation to the physical world, it is surely rash to assume this without giving the matter very careful consideration.

My main purpose in the present chapter has been to dispel any notion that we can simply ignore the modern scientific understanding of spacetime, on the grounds that it applies only within the context of rarefied scientific theorizing. On the contrary, I have argued that the scientific understanding is what we get when we bring the ordinary understanding to its fullest logical development in the light of relevant empirical observations (such as the observations that establish the frame-independence of the speed of light). In the chapters that follow, we consider how the scientific understanding constrains the project of assigning the mind its proper place in nature.

3

Phenomenal duration, succession, and change

And how are we to deal with the appearance itself? If we reduce time and change to appearance, must it not be to an appearance which changes and which is in time, and is not time, then, shown to be real after all? This is doubtless a serious question, but I hope to show hereafter that it can be answered in a satisfactory way.

J. Ellis McTaggart, *The Unreality of Time*

3.1 Introduction

In this chapter, I argue that the fact that we have conscious experiences characterized by phenomenal duration, succession, and change does not give us any good reason to think that our conscious experiences occur in physical time. If successful, the argument meets the first of the principal challenges to radical phenomenalism enumerated at the beginning of this book.

Most philosophers of mind, and probably most people, consider it to be just obvious that conscious experience is a temporal phenomenon, unfolding in and changing over time. As J.J. Valberg puts it,

> it seems obvious that, whatever exactly we are talking about when we talk about 'how things are in our experience', we are not talking about something atemporal. On the contrary, it seems obvious that we are talking about something which can change (which is in fact changing all the time) and is thus temporal.[1]

Some philosophers go farther, holding that it is not merely obvious that conscious experience occurs in (and changes over) physical time, but even a conceptual or a priori truth that it does so. According to John Stuart Mill, for example,

[1] Valberg (1992), 37.

Sensations exist before and after one another. This is as much a primordial fact as sensation itself; it is a feature always present in sensation, and we have the strongest ground that can ever be had for regarding it as ultimate, because every genesis we assign to any other fact of perception or thought, includes it as a condition.[2]

This recalls Kant's remark that

[t]he idea of time does not originate in, but is presupposed by the senses. Whether things falling under sense-perception be simultaneous or in line of succession cannot be represented but by the idea of time; nor does succession beget the concept of time; it appeals to it. Hence the notion of time, though acquired by experience, is badly defined by a series of actual things existing one after another, for what the word *after* means I understand only by the previous concept of time.[3]

While it may seem obvious or inevitable that consciousness changes over time, this apparent obviousness or inevitability is, I think, an illusion.

From the fact that I have an experience as of F, it does not, in general, follow that I have an F experience. An experience as of a loud noise is not itself loud; any "loudness" that we attribute to the experience is a phenomenal loudness distinct from the physical loudness of, say, an explosion. An experience as of a colorful scarf need not be colorful, an experience as of a cold wind need not be cold, an experience as of a sour grape need not be sour. These experiences are *phenomenally* colorful, cold, and sour, but they are not (or anyway needn't be) colorful or cold or sour in the same sense as their presumed objects (scarves, winds, grapes).

If, as Kant and Mill maintain, an experience as of a succession of events must itself be characterized by temporal succession (an objective succession of qualia-instantiations in physical time), then in this respect phenomenal succession is exceptional—likewise for phenomenal duration and phenomenal change. The Kantian or Millian doctrine implies that unlike phenomenal color, phenomenal loudness, etc., phenomenal succession cannot exist except in the form of an experience that exhibits its objective counterpart (succession in physical time). According to Kant and Mill, phenomenal succession presupposes physical succession (succession in physical time), even though phenomenal noise does not presuppose physical noise.

There is, then, no direct a priori argument for the claim that human experience involves an element of duration, succession, or change.

Still, it has seemed to many philosophers that there *is* a compelling argument for this claim, which is that we have direct introspective evidence that our experiences have these temporal features. According to Locke,

[2] Mill (1865/1979), 198–9. [3] Kant (1770/1894), 59.

> [i]t is evident to anyone who will but observe what passes in his own mind, that there is a train of ideas which constantly succeed one another in his understanding as long as he is awake.[4]

Ian Phillips concurs with Locke on this point, maintaining that

> we cannot make sense of the idea that experience systematically seems to one's rational introspective reflection to possess a certain temporal ordering, when it is not in fact genuinely so ordered.[5]

But, again: why should rational introspective reflection give us any more reason to think that experience possesses a certain temporal ordering—an objective ordering in physical time—than it gives us to think that experience possesses a certain physical color, flavor, or odor?

According to Matthew Soteriou and Oliver Rashbrook, the answer to this question is that we have introspective grounds for believing that the physical duration of a conscious experience is the same as the physical duration of that of which it is an experience. They argue that since "experience *is as it seems,*" and since our experiences generally seem to us to endure for just as long as the events, states, or processes of which they are experiences, it follows that our experiences have the same durations as their respective objects (the events, states, or processes of which they are experiences).[6]

However, this pushes the principle that "experience is as it seems" too far. The principle is plausible when it comes to the intrinsic phenomenal properties of an experience: if an experience seems to the one who has it to be phenomenally red, then it is phenomenally red (at least, if the experience seems phenomenally red to the subject under ideal circumstances of introspection). But when it comes to the *extrinsic* properties that an experience has by virtue of relating to non-experiential things, the principle is not plausible. In particular, the fact (assuming it to be a fact) that an experience generally seems to the person who has it to have the same duration as its object gives us no more reason to think that the experience has the same duration as its object than the fact that the pain of a stomach cramp generally seems to the person who has it to occur in his stomach gives us a reason to think that the pain occurs in his stomach. Indeed, one could argue that in a dream, an

[4] Locke (1690/1979), 122.

[5] Phillips (2010), 183.

[6] In effect, Soteriou and Rashbrook endorse what Izchak Miller calls the "Principle of Presentational Concurrence"; see Miller (1984), 107, quoted in Dainton (2006), 134. For Soteriou's and Rashbrook's views, see Soteriou (2010), 227, and Rashbrook (2013), 588–601, 608–9. Obviously, one would have to do some work to extend the proposed principle to non-veridical experiences that have no actual objects, such as dreams.

experience typically does *not* have the same duration as its object (or would-be object), but usually a much shorter duration.

That we have direct introspective evidence of *phenomenal* duration, succession, and change—direct introspective evidence, that is, of the existence of *appearances* of things happening (or, as the case may be, not happening) over time—is, I think, undeniable. One might as well deny the reality of phenomenal shape, color, or odor. The purpose of this chapter is to argue that this does not count as evidence for the existence of anything that endures, succeeds, or changes in physical time. The argument is this:

A1. Each of my introspectible conscious experiences either: (i) fails to instantiate qualia in temporal succession; or (ii) comprises experiences each of which fails to instantiate qualia in temporal succession.
A2. Each of my experiences is an experience as of temporal succession.
A3. So, some of my experiences as of succession do not instantiate qualia in temporal succession (from A1 and A2).
A4. If some of my experiences as of succession do not involve a successive instantiation of qualia, then introspection alone gives me no basis for thinking that any aspect of my mental life is characterized by objective temporal succession.
A5. So, introspection gives me no basis for thinking that any aspect of my mental life is characterized by objective temporal succession (from A3 and A4).

This argument has three underived premises: (A1) that human experience is, as Leibniz would put it, a "well-founded phenomenon"; (A2) that phenomenal succession is ubiquitous in human experience; and (A4) that I have no introspective basis for thinking that qualia get instantiated successively if I have appearances of succession that involve no succession of appearances (no successive instantiation of qualia). I devote the remainder of this chapter to substantiating these claims.

3.2 The well-foundedness of experience (A1)

Define a "diachronically simple" experience as an experience that does not instantiate qualia in temporal succession (and so does not have other, temporally successive experiences as parts or phases). A1 says, in effect, that every human experience is diachronically simple, or else consists entirely of diachronically simple experiences; it says that human experience is a well-founded phenomenon. (To deny that human experience is a well-founded phenomenon is to say that it consists of phenomenal "gunk.")

There is actually a nuance here that bears mention. Formulated more precisely, the claim that experience is well founded is the claim that every human experience that is constituted by successive experiences *belonging to the same mind as the constituted experience* is constituted by experiences that do not themselves consist of successions of experiences belonging to that mind. When I say that human experience is well founded, I do not mean to deny the possibility of the Visual Brain or any similar arrangement of experiences. The well-foundedness of human experience is compatible with the (bizarre) claim that each of my experiences is composed out of some sequence of other beings' experiences, and that each of those beings' experiences is composed out of successive experiences belonging to yet other beings, and so forth without end.

I listen to the wind rushing through a stand of trees on an autumn evening. If human experience is not well founded, then the experience I have, call it *E*, and each of *E*'s constituents, has an infinite number of successive constituent experiences. This follows from the fact that phenomenal constituency is transitive: if *E* has e_1 as a constituent, and e_1 has e_{1_1} and e_{1_2} as constituents, then *E* has e_{1_1} and e_{1_2} as constituents.[7]

But if *E* (or any other conscious experience) were phenomenally constituted by an infinite number of successive experiences, there would not be any lower bound on the constituent experiences' phenomenal durations. It would have to be that for any experience of mine as of a given time interval, I could have an experience as of a shorter interval. Otherwise, *E* would be phenomenally constituted by an infinite number of successive experiences, each having no less than a certain minimum phenomenal duration—in which case, *E* would be an appearance of an infinitely prolonged state of affairs, whereas I have no such experience.

So my experience *E* consists of an infinite number of successive experiences only if there is no lower bound on the phenomenal duration of my experiences.

But there *is* a lower bound on the phenomenal duration of my experiences. If there weren't, then there would be nothing to prevent me from having a visual experience as of a light shining for one billionth of a second, or as of one flame

[7] I do not put this forward as a general claim about the nature of constitution. Possibly there are intransitive forms of constituency. But in the sense of "constitute" in which several experiences may be said to constitute a greater experience, constituency is transitive. If the Visual Brain has a complex auditory experience, this experience may have simpler auditory experiences as constituents, in the relevant sense. If these simpler experiences are themselves constituted by visual experiences (visual experiences playing the role of neurons in the Visual Brain), this will be in a different sense of "constituted." The statement that the complex auditory experience is constituted by simpler auditory experiences without being constituted by those simpler experiences' constituents will then be true, but only because it involves an equivocal use of "constituted." Here, we are interested only in the first sense of the word, in which the simpler auditory experiences are said to constitute the complex auditory experience. And in this sense, constitution is transitive.

flickering out one billionth of a second after another flame. Likewise, if there were no lower bound on the phenomenal duration of my experiences, then I could have a "nano-experience" as of an instrument playing for one billionth of a second, or an auditory experience consisting of 999,999,999 successive nano-experiences as of a flute accompanying a violin, and a billionth nano-experience as of a flute playing solo, the entire sequence being an experience as of a flute playing for one nanosecond longer than a violin. The fact that experiences like these are not possible for me shows that there is a lower bound on the phenomenal duration of my experiences.

Could one object that I do have experiences as of nano-events, but that I am simply unable to introspect them? Or that I have experiences as of durational nano-differences that I cannot help but mistake for experiences as of flames going out simultaneously (or sounds ending simultaneously, or whatever)?[8]

Well, it is possible that we sometimes make mistakes about the phenomenal character of our own experiences. But on the present suggestion, we are constantly and massively deceived about the phenomenal characters of our experiences, even when we introspect them under ideal conditions. If there is no lower bound on the phenomenal duration of my experiences, then when I have an experience as of a bird call, I also have a phenomenally distinct experience as of an arbitrarily briefer portion of a bird call, as well as a phenomenally distinct experience as of a portion of a bird call arbitrarily briefer than *that,* etc. But no matter how hard I try to discern these alleged experiences, and no matter how ideal the circumstances are under which I atttempt to introspect them, I do not find them in me. This shows that I do not have such experiences at all.[9]

Indeed, not only does ideal introspection fail to reveal the existence of such experiences in me: I cannot even *conceive* of having such experiences. Suggesting that I have had an experience as of one sound ending a nanosecond earlier than another is not like suggesting that I have had an experience as of a ten-headed chicken. It is like suggesting that I have had an experience as of a ten-dimensional solid.

I conclude that the existence of a lower bound on the phenomenal duration of my experiences is indeed known from introspection, and that there is no plausible way to deny this by appeal to introspective ignorance or error.

To bring this section to a close, let me summarize the argument for the claim that human experience is well founded (i.e. A1): Human experience is well founded,

[8] See Graff (2001), 927–8.

[9] Recall the arguments of Chapter 1, Section 4. On the question of a lower bound on the phenomenal duration of human experiences, I find myself in complete agreement with Ian Phillips: see esp. Phillips (2011), 813; see also Wright (1975), 346.

unless we have experiences that are infinitely divisible into further, successive experiences. But we have experiences that are so divisible only if there is no lower bound on the phenomenal duration of our experiences. But there *is* a lower bound. So we do *not* have experiences that are infinitely divisible into further, successive experiences. So human experience is a well-founded phenomenon.

3.3 The ubiquity of phenomenal succession (A2)

Every experience that I have is an experience as of persistence, succession, or change: an experience as of things happening, or, as the case may be, not happening, over some (generally quite brief) period of time. In this respect, I take myself to be no different from any other human being.[10]

J.M.E. McTaggart, best known for denying the reality of time, nevertheless recognizes the reality of conscious *appearances* of time—experiences *as of* events occurring in succession, objects undergoing changes, etc. Indeed, McTaggart considers phenomenal temporality to be a pervasive feature of human experience, remarking that we never perceive things except as existing in time.[11] Call this

McTaggart's Principle: A typical human experience is an experience as of persistence, succession, or change.

McTaggart's Principle is a claim about the nature of typical human experience. Maybe there are beings who have experiences that are not as of things happening (or not happening) over time; maybe even a human being could have such an experience (perhaps due to some neurological pathology or the influence of a powerful drug). All that McTaggart's Principle says is that if such experiences exist in the human sphere, they are atypical and rare.

McTaggart's principle receives strong support from introspection. If I try to imagine a state of affairs involving no persistence, succession, or change, I fail.

[10] Phenomenal succession is not to be confused with a supposed quality of "phenomenal passage" characterizing putative experiences as of events changing from future to present, or from present to past. Talk of "appearances of passage," or of "an experience as of events passing," is, I believe, confused—on this point I find myself in agreement with Hoerl (2009), 14–15. I am never tempted to describe any of my experiences as an experience as of something happening in the past (or in the future). Maybe such experiences are possible, but I have never had one. Assuming that other people resemble me in this respect, it cannot be any such experience that people have in mind when they speak of "phenomenal passage," or "a subjective appearance of the passage of events." Probably what they are getting at is the fact that all our appearances are appearances of temporal *succession*—of events unfolding in physical time, or things happening (or, as the case may be, not happening) over some temporal interval. Such an appearance is an experience as of some events occurring earlier than or later than others, but it is not (for me at least) an experience as of some events occurring in the past, or in the future.

[11] See McTaggart (1908), 470.

I cannot even conceive of what it would be like to have an experience that was not an experience as of things happening (or not happening) over time, and this more or less proves that I have no such experience.

There are, I think, only three reasons why one might hesitate to accept McTaggart's principle, none of which survives critical reflection.

First, you might suggest that it is an illusion that we have conscious experiences as of things happening over time. You might suggest that in reality, we only ever have experiences as of momentary and unchanging states of affairs, but that these experiences occur in such rapid succession that they give rise to an illusion of phenomenal persistence, succession, or change. One might try to support this suggestion with an analogy to watching a movie, where the images are projected onto the screen one at a time, but so quickly that you cannot distinguish one frame from the next.

But how could a rapid sequence of phenomenally durationless and unchanging experiences create an illusion of phenomenal change without thereby creating a phenomenally changing experience? I cannot even conceive of an experience that I would be inclined to describe as an illusory appearance of phenomenal change. I can easily imagine an illusory appearance of *physical* change; the waterfall illusion is one example. But the experiences I have when subject to the waterfall illusion are experiences as of change.

The cinematic analogy is not apt. At the movies, the successive images on the screen do not occur as part of my mental life, but only as external causes or correlates of mental events. It would be more appropriate to consider things from the point of view of the movie screen, if the movie screen had a point of view, and if the images projected onto it belonged to it as conscious experiences. It can be an illusion that the movie screen is covered with moving images, but it cannot be an illusion (to me) that I have an *experience* as of moving images. There is just no place here to separate appearance from reality.

The second reason you might have reservations about McTaggart's Principle is that you think we often do have experiences as of durationless states of affairs. One might argue that any perception of an abrupt change involves an appearance of a durationless event; for example, a durationless transition from darkness to light, or from silence to noise, or stillness to motion.[12]

Let's grant, for the sake of argument, that we do have experiences as of durationless and unchanging events. Still, each of these experiences is *also* an appearance

[12] You could also argue that we have experiences as of spatially unextended things. Consider a visual perception of a square made up of four different-colored smaller squares. One might argue that this involves an appearance of a spatially unextended entity: the unextended point at which the four smaller squares meet at the center of the larger square that they compose.

of things happening over time. I cannot have an experience that is *only* as of a transition from darkness to light: it must also be an appearance of extended (if brief) periods of darkness and light; I cannot have an experience that is *only* as of a transition from stillness to motion: it must also be an appearance of extended (if brief) periods of stillness and motion. Any appearance of the durationless that I may have is also an appearance of the temporally extended.[13]

Consider the Weierstrass function. This function is defined at every point within a continuous interval, but nowhere differentiable (and so everywhere "jagged") within that interval. If you had an infinitely detailed graph of the function, and an infinitely powerful microscope, the graph would appear jagged at all levels of magnification. Zooming in on one part of the graph at one level of magnification, you would see what appeared to be a series of angles or jags, each composed of smooth slopes joined at a point. Zooming in further, these "smooth" slopes would resolve themselves into further series of jags, and so forth ad infinitum (see Fig. 3.1, where the graph-sequence zooms in on the peak of the function at $x = 0$).

The Weierstrass function has no counterpart in human experience. We cannot (or anyway do not) have an experience as of a geometrical structure that is everywhere defined but nowhere differentiable; that is, so to speak, all peaks and troughs and no slopes. Human beings cannot visualize a peak (or trough) except by visualizing smooth slopes to bound it. If they could, Weierstrass's discovery of the function wouldn't have come as such a surprise. By the same token, we cannot have an experience as of a process described by the Weierstrass function: a process that consists of nothing but instantaneous transitions from one state to another, with no temporally extended bounding states. For us, to be conscious is to be subject to an appearance, or appearances, of things changing or staying the same. Yet, if it were possible for us to have experiences that were not experiences as of things changing or staying the same, there should be nothing to prevent us from having an experience as of such a "jagged" process. The fact that we cannot have such an experience is proof that each of our experiences is, inter alia, an experience as of duration, succession, or change.

McTaggart's Principle says that all of our appearances are appearances of duration. It does not say that all of our appearances are *not* appearances of durationless events. The principle only requires us to say that none of our experiences is an experience *only* as of something durationless, and this is something that experiences as of sudden changes do not cast in doubt.

The third and final reason why one might hesitate to accept McTaggart's Principle is for fear that it entails a kind of vicious phenomenological regress. The

[13] Soteriou has made this point in Soteriou (2010), 226.

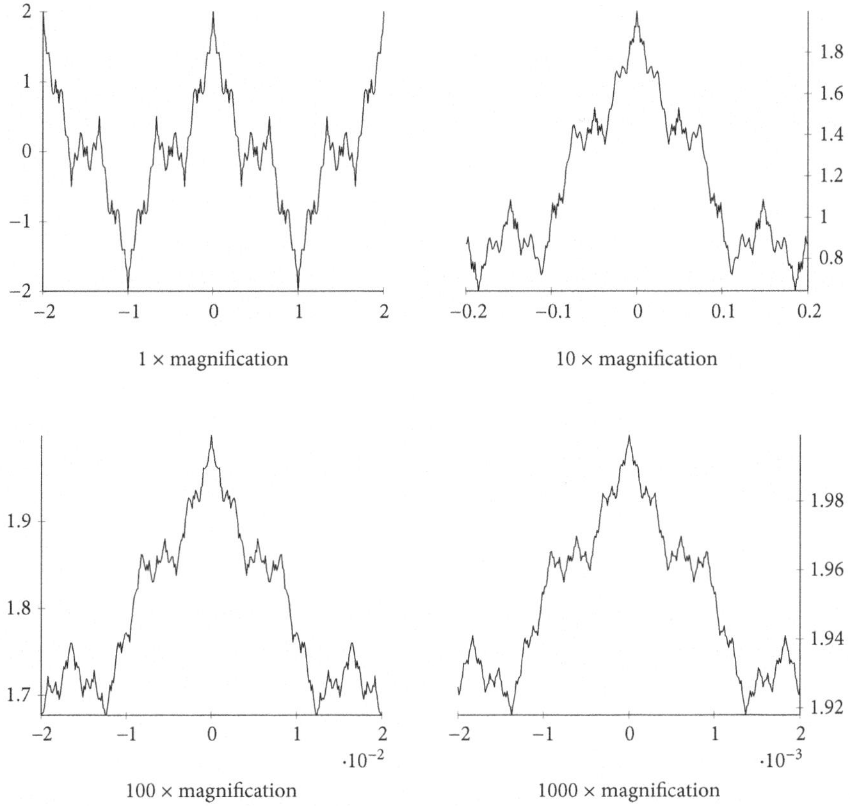

Figure 3.1 Weierstrass function: bumpy all the way down.

principle implies that every experience is an experience as of an enduring object, event, or state of affairs. So, take an experience, *E*, as of a state of affairs with some duration *d*. Since *E* is an experience as of something with duration *d*, mustn't it also be an experience as of some state of affairs with duration $d/2$? And if *E* is an experience as of a state of affairs with duration $d/2$, mustn't it also be an experience as of some state of affairs with duration $d/4$? In short, doesn't McTaggart's Principle entail that every one of my experiences is, among other things, an experience as of arbitrarily brief states of affairs? And doesn't that conflict with the conclusion drawn above, that there is a lower bound on the phenomenal duration of a human experience?

The solution to this paradox is that an experience as of a temporally extended state of affairs need not also be an experience as of a briefer state of affairs (such as a state of affairs half as long as the one as of which the experience is an experience). For there may be a distinctive phenomenal quality common to all typical experiences of, say, an event that lasts ten milliseconds, without there being any distinctive phenomenal quality common to all typical experiences of events lasting five milliseconds. An experience whose phenomenal duration is at the lower bound of what is possible for me is an experience like that. Supposing otherwise would be like inferring that we have experiences as of zero-dimensional objects from the fact that we have experiences as of three-dimensional objects.[14]

I can have an experience as of a star-studded sky, and I can have an experience as of a sky containing just a single star; from this, it does not follow that I can have an experience as of a sky that contains forty-two stars (although of course I can have an experience *of* a sky containing forty-two stars). No more does it follow from the fact that I can have an experience as of an enduring state and an experience as of an instantaneous state, that I can have an experience as of a state that lasts for one millisecond.

3.4 The general case (A4)

McTaggart's Principle and the well-foundedness of human experience jointly entail that many of our experiences (our diachronically simple ones) are appearances of temporal succession that do not involve any temporal succession of experiences—any successive occurrence of phenomenal states, or any successive instantiation of qualia, in physical time. This gives us premise A2 of the main argument, which, combined with A1 (which we have already established), entails A3.

The third and final step of the main argument, A4, states that if I have even one appearance of succession that does not involve a successive instantiation of qualia, then introspection gives me no basis for thinking that *any* aspect of my mental life is characterized by objective temporal succession.

The rationale for this is straightforward. If introspection gives me a basis for making a judgment of objective temporal succession—for judging that one event happens after another, or one state of affairs obtains after another, in physical time—it gives me that basis by revealing the existence of conscious experiences as of temporal succession and change. Every such experience equally supports, or equally fails to support, a judgment of objective temporal succession (a judgment to the effect that the introspected experience successively instantiates various qualia in physical time). I may have appearances of change that *resist* construal as

[14] Of course, I have conceded that we can, perhaps, make sense of the suggestion that we have experiences as of zero-dimensional things (e.g. cross-cutting color boundaries); but our having such experiences is not a necessary condition for our having experiences as of three-dimensional objects.

diachronically complex experiences, due to the fact that in order to be diachronically complex, they would have to have as constituents experiences as of events briefer than any that can appear to me. But none of my appearances *demands* to be construed as a diachronically complex experience, since, as far as introspection can reveal, the qualia that characterize any of them might belong to a single, diachronically simple experience.

Here, one might object that some appearances of change *do* demand to be construed as diachronically complex. One might suggest that there is an important phenomenal difference between experiences as of brief changes, and experiences as of changes that play out over a period of minutes or hours, and that this difference is best explained by construing the latter experiences, but not the former, as diachronically complex.

This objection is based on a confusion. In ordinary human experience, there is no such thing as an experience as of a change that plays out over minutes or hours. There is, to be sure, such a thing as a *stream of consciousness* that contains many experiences as of change, where these experiences are best collectively interpreted as perceptions of events unfolding over a period of hours. But such a stream is not a single experience—not even a single complex experience. In the morning, I have an experience as of the Sun peeking above the horizon, and in the evening, I have an experience as of the Sun disappearing below the horizon. But I have no experience as of the Sun rising and then setting. Such an experience is possible; I might have it if I were standing on the Arctic Circle at Winter Solstice. But my actual stream of consciousness neither is nor contains such an experience. That is why in order to know that my actual stream of consciousness includes an appearance of the Sun rising as well as an appearance of the Sun setting, I must draw on more than introspection: I must rely on memory.

For all that has been argued so far, we may have to locate conscious experience in physical time in order to account for the fact that our experiences typically belong to extended streams of consciousness. That is a question we take up in Chapter 5. The purpose of the present chapter is only to show that there is nothing in the phenomenal character of any *experience* as of duration, succession, or change that compels us to locate consciousness in physical time. Confining our attention to experiences, the point stands that if any of my appearances of succession fails to give me an introspective basis for judging that conscious experience includes objectively successive elements (sensations, phenomenal states, qualia instantiations, or whatever), then *none* of my conscious experiences gives me an introspective basis for such a judgment. And if none of my conscious experiences gives me an introspective basis for making such a judgment, then introspection gives me no basis for thinking that conscious experience involves any kind of succession in physical time. This gives us A4, and concludes the main argument of this chapter.

3.5 Conclusion

I have argued that the fact that our experiences exhibit phenomenal duration, succession, and change gives us no reason to think that they exhibit objective duration, succession, or change (duration, etc. in physical time). *Pace* Kant and Mill, it is not an a priori or conceptual truth that phenomenal duration, succession, and change presuppose their objective counterparts. *Pace* Locke and his followers, introspection yields no compelling evidence of objective temporal duration, succession, or change in our experience—no evidence of successive sensations, or successive developments of a single sensation; no evidence of any sequential instantiation of qualia; etc. The phenomenal duration of an experience is not only independent of any physical duration that the experience might be said to have, but even, as far as the evidence of introspection can establish, independent of whether the experience has any physical duration at all. Evidence of *phenomenal* temporal extension is not evidence of extension in physical time. This overcomes the first objection to radical phenomenalism listed at the beginning of this book.

But the evidence of introspection includes more than evidence of phenomenal duration, succession, and change. Introspection also reveals experiences as of simultaneous events. Maybe phenomenal simultaneity, unlike phenomenal succession, *does* somehow require us to locate consciousness in time. We consider this possibility in Chapter 4.

4

Phenomenal simultaneity

> Two events which are simultaneous in my experience may be spatially separate in psychical space, e.g. when I see two stars at once. But in physical space these two events are not separated, and indeed they occur in the same place in space-time. Thus in this respect relativity theory has complicated the relation between perception and physics.
>
> The problem is one whose importance and even existence has been concealed by the unfortunate separation of different studies which prevails throughout the civilized world . . . Philosophers, it is to be hoped, will come to recognize that they cannot achieve any solid success in such problems without some slight knowledge of logic, mathematics, and physics; meanwhile, for want of students with the necessary equipment, this vital problem remains unattempted and unknown.
>
> Bertrand Russell, *The World of Physics and the World of Sense*

4.1 Introduction

In the paragraphs quoted above, Russell suggests that the phenomenal properties that characterize any moment of one's conscious mental life belong to an experience that is confined to a single point of spacetime. He gives the example of seeing two stars: this is an experience as of spatially separated points of light shining simultaneously, just as all, or nearly all, of our visual experiences are appearances of objects simultaneously distributed in space. But according to Russell, the experience occurs in physical spacetime as a single point-event.

Russell does not make it clear whether he means to suggest that *all* of our conscious experiences are point-events in spacetime, or just some of them. Presumably he would say that at least all of our diachronically simple experiences are point-events, since a diachronically simple experience involves no successive instantiation of qualia. Let us call this:

Russell's Thesis: diachronically simple experiences are point-events.

Our discussion in this chapter will revolve around this thesis.

In Section 4.2, I give what I take to be Russell's argument for the thesis. In Section 4.3, we consider a challenge to this argument. The challenge takes the form of a theory by which there is no such thing as an absolutely simultaneous instantiation of qualia—no moment when consciousness is wholly present—because every quale gets instantiated (if at all) by a whole sequence of events (e.g. a whole sequence of synapse-firings). I call this theory "serialism" about conscious experience.

As we shall see, serialism violates a highly plausible (and arguably a priori) thesis about conscious experience, to the effect that a conscious subject's experience is, in a certain sense, *unified*. On the other hand, and as we shall also see, rejecting serialism in favor of Russell's view leads to a peculiar, though not obviously indefensible, form of dualism.

So, by the end of the chapter, we'll find ourselves confronted with an inescapable choice among three options: (1) accept a Russellian dualism that treats our most basic conscious experiences as non-physical point-events; (2) accept a serialist theory that entails the possibility of a dubious kind of phenomenological disunity; and (3) deny that conscious experience exists in spacetime (i.e. accept Poincaré's phenomenological atemporalism).

4.2 Russell's Thesis

Russell claims that at least very many of our conscious experiences, including presumably all of our diachronically simple experiences, occur in physical spacetime as singular point-events. What led him to this conclusion?[1]

He gives no explicit argument, but it is not hard to guess what he had in mind. At any given moment of consciousness, one's conscious experience has its various phenomenal properties absolutely simultaneously; any conscious experience that has its phenomenal properties absolutely simultaneously is a point-event (confined to a single point of relativistic spacetime); therefore, at any given moment that one is conscious, one's conscious experience is confined to a single point of spacetime. The core of this argument is as follows:

B1. In my conscious mental life, it often happens that multiple qualia get instantiated absolutely simultaneously.

[1] Charles Peirce makes a similar claim, though on grounds unrelated to the then-unknown relativistic conception of spacetime: see Peirce (1892), 536.

B2. An absolutely simultaneous instantiation of qualia in my mental life is an experience that is confined to a single point of spacetime.

B3. So, I have conscious experiences, each of which is confined to a single point of spacetime (i.e. each of which is a point-event).

Given that our diachronically simple experiences are point-events if any of our experiences are, it is a short step from this conclusion to Russell's Thesis.

There are two erroneous reactions to the foregoing argument that we should forestall at the outset.

The first error is to reject B2 on the grounds that conscious experience occurs in physical time, but not in physical space, and therefore not in spacetime (and not, therefore, in relativistic spacetime). If correct, this means that absolutely simultaneous experiences (or qualia instantiations), unlike absolutely simultaneous physical events, need not be spatially (or spatiotemporally) co-located.[2]

But an event (whether an experience or anything else) cannot occur in physical time except by occurring in physical spacetime. A description of events purely in terms of physical time is an underdescription of events, not only because it does not tell us how the events relate to one another in space, but also because it does not tell us which events occur *absolutely* before, after, or simultaneously with which others. The statement that event E_1 occurs at the same time as event E_2 does not convey whether the simultaneity of E_1 and E_2 is absolute or merely relative; for that, we need to know whether E_1 and E_2 occurred at the same place (in addition to the same time). Likewise, facts about events' times of occurrence alone do not determine whether any two events are timelike, lightlike, or spacelike separated; for that, we need information about the events' spatial locations.

This means that if experiences do not occur in space, then there is no fact of the matter as to how any experience relates in time to anything else. To say that experiences occur in physical time but not in physical space would therefore be to claim, implicitly, that there was no fact of the matter concerning the temporal relationships between our conscious experiences and ordinary physical events. But this would be as much as to say that conscious experience does not occur in physical time at all.

The second error is to reject B2 on the grounds that the relativity of simultaneity applies only to physical events, but not to conscious experiences, even assuming that conscious experiences occur in physical spacetime. Maybe conscious experiences occur in the same spacetime as physical events, but in a non-relativistic way that permits spatially separated experiences to occur absolutely simultaneously.

[2] See, for example, McGinn (1995).

But if phenomenal events occur in the same spacetime as physical events, it must be possible for a phenomenal event to occur at the same location in spacetime as a physical event. This is particularly so in view of the fact that physical events occur throughout spacetime (e.g. as components of the cosmic microwave background). Consider two spatially separated, and supposedly absolutely simultaneous, phenomenal events, A_1 and A_2, and a pair of physical events, B_1 and B_2, with which the phenomenal events respectively coincide. Then A_1 is absolutely simultaneous with B_1, and A_2 with B_2. But then it must also be the case that B_1 is absolutely simultaneous with B_2, since A_1 is absolutely simultaneous with A_2, and absolute simultaneity is transitive. The upshot is that if spatially separated phenomenal events can be absolutely simultaneous, spatially separated *physical* events can be absolutely simultaneous too. Since separate physical events cannot be absolutely simultaneous, neither can separate phenomenal events.

Putting these confused objections behind us, let us look again at the premises of Russell's argument.

B2 is actually two claims in one: that an absolutely simultaneous instantiation of qualia in my mental life is an experience, and that it is an experience that is confined to a single point of spacetime.

The former claim is just a conceptual truth about consciousness. It can be expressed as a conditional: if some qualia that characterize my conscious mental life are instantiated simultaneously, then the moment at which those qualia get instantiated contains a conscious experience that those qualia characterize. One should accept this conditional, even if one rejects its antecedent.[3]

The second claim contained in B2 is that an experience whose qualia get instantiated absolutely simultaneously is a point-event in spacetime. As Russell points out, this is a direct consequence of the fact that in relativistic spacetime, absolute simultaneity entails spatial and temporal co-location.

What about B1? Russell takes this to be evident from introspection. According to him, when I look at a vase that holds a yellow daffodil and a red rose, I have an experience that simultaneously instantiates various qualia—various phenomenal shapes, sizes, colors, positions, etc. My experience's simultaneous possession of these phenomenal features is moreover *absolute,* in the sense that the experience instantiates the properties simultaneously according to every complete and accurate description of the world. If you were to describe me as having first an experience as of the daffodil, and only then an experience as of the rose, you would misdescribe my experience of the flowers. So when I see the flowers, I have a conscious visual appearance that is simultaneously characterized by (at least) two

[3] The conditional is equivalent to the "unity thesis" discussed in Section 4.3.

color-qualia, two shape-qualia, etc. B1 simply says that I often have experiences like this—experiences characterized by simultaneously instantiated qualia.

But B1 may not be as unassailable as Russell supposes. Certainly, introspection reveals that when I see the flowers, I have a visual experience that instantiates various phenomenal properties. I also know from introspection (or a combination of introspection and memory) that my experience of the two flowers does not instantiate these qualia in temporal succession (e.g. first the daffodil, then the rose; or first the shape of the rose, and then the size of the rose). But from the fact that a given experience instantiates two qualia without instantiating them in temporal succession, it does not automatically follow that it instantiates those qualia simultaneously, even assuming that the experience occurs in time. For it may be suggested that my visual experience of the flowers is an objective temporal sequence of non-experiential events, and that it is this whole sequence of events that instantiates the color-, shape-, size-, and position-qualia, somewhat as a whole sequence of events instantiates the property of being a golf tournament.

More generally, one might hold that *every* conscious experience is a sequence of non-experiential events (brain-events, say), where it is the whole sequence, rather than any individual event in it, that instantiates the qualia that characterize the experience. Let us call this suggestion, which does not seem to have occurred to Russell, "serialism" about conscious experience.[4]

4.3 The unity of the Now

According to Russell, there are many moments at which conscious experience is wholly present; according to the serialist, there is no moment at which consciousness is wholly present. When I look at a vase that holds a yellow daffodil and a

[4] For sympathetic statements of the serialist position, see Soteriou (2007), 551–5, Soteriou (2010), 235, and Rashbrook (2013), 605–7. Brian O'Shaughnessy and Galen Strawson also advocate serialism, on the somewhat radical grounds that only processes exist; see O'Shaughnessy (2000), 42–4 and Strawson (2008), 159–66. Strawson claims to be a physicalist, but one could be a serialist without being a physicalist: one could think of a conscious experience as a sequence of physical events that instantiates a non-physical quale.

Serialism is a claim about the nature of individual *experiences*, such as an individual visual appearance of a bird in flight. Not all perceptual activities take the form of individual experiences; for example, watching a bird circle overhead for five minutes does not. Whether a "watching" like this reduces to a sequence of non-watchings is an interesting question in its own right, but not strictly relevant to the assessment of serialism; for a detailed discussion of watching, see Crowther (2009).

Serialism is *not* the claim that a conscious experience is essentially a *culmination* of successive non-experiential events. Such a view would imply, wrongly, that a conscious experience logically depended for its existence on the occurrence of prior (non-experiential) events. In that case, one could refute skepticism about the existence of past physical events simply by referring to the fact that we are conscious.

red rose, I have an experience as of the simultaneous existence of a yellow flower and a red flower. That is not in dispute. The question is whether this experience is simultaneously an experience as of a yellow flower and an experience as of a red flower. Russell says it is, the serialist says it is not.

According to the serialist, my visual experience of the flowers is some sequence of non-experiential events, or rather: it is the instantiation of some quale or qualia by a sequence of non-experiential events. More generally, serialism implies that it is possible to parse my life—or any life—into intervals of time, none of which contains any conscious experience. Serialism implies that every conscious being is, at some level of description, never conscious. Of course, the serialist will say that such a level of description fails to capture everything about the conscious being. He will say that being conscious is like walking, or oscillating. At some level of description, an oscillating being never oscillates; *mutatis mutandis* for a conscious being and having conscious experiences. Or so says the serialist.[5]

At first, Russell and the serialist seem to occupy equally coherent positions. And if the positions are equally coherent, it seems reasonable to favor serialism—at least, if we want to think of consciousness as existing in physical time. For serialism allows all conscious experiences to have spatiotemporal extent, which makes it much easier than it otherwise would be to relate our experiences to the other contents of spacetime (such as brain processes).

However, the coherence of the serialist position may be an illusion. For serialism implies the possibility of a seemingly impossible kind of phenomenal disunity—an inconceivable alienation of the conscious mind from itself.

In the *Monadology*, Leibniz writes that "the *perception* . . . is *inexplicable in terms of mechanical reasons,* that is, through shapes and motions." He continues:

> If we imagine that there is a machine whose structure makes it think, sense, and have perceptions, we could conceive it enlarged, keeping the same proportions, so that we could enter into it, as one enters into a mill. When inspecting its interior, we will only find parts that push one another, and we will never find anything to explain a perception. And so, we should seek perception in the simple substance and not in the composite or in the machine.[6]

[5] Why must the serialist say that the experience is a sequence of non-experiential events (or the instantiation of a quale by such a sequence)? Because, as we learned in Chapter 3, not every experience is a temporal sequence of experiences. So even if I do have a diachronically extended experience when I see the flowers—an experience that has briefer, successive experiences as parts or phases—I also have an experience that does *not* have successive experiences as parts. The discussion then proceeds with reference to that experience, which must, according to the serialist theory, and given the diachronically simple nature of the experience, consist of some sequence of non-experiential events.

[6] Leibniz (1714/1989a), 215.

Leibniz presents his thought-experiment as part of an argument against functionalism; indeed, he comes across as favoring a pointillistic dualism (although later we shall see that Leibniz is actually more an atemporalist than a pointillist). That argument might be too hasty. A physicalist might reply that if you were to perceive the workings of the sentient mill you *would* perceive conscious sensations. You would not perceive them *as* sensations, but the things you perceive *as* operations of the mill might *be* sensations for all that.

But even if we can make sense of the idea that what Leibniz perceives as the shifting of gears, pushing of rods, etc. are in fact conscious experiences, the scenario loses its intelligibility if we stipulate that the workings that Leibniz perceives are conscious experiences that Leibniz himself is having. This is not just because we can't imagine Leibniz walking around inside his own head—as we shall see momentarily, this is a complication that some minor modification to the thought-experiment easily overcomes—but because the stipulation implies an incoherent disunity in Leibniz's conscious experience.[7]

Let's imagine a creature that has detailed visual experiences of neural processes that constitute or underlie its own auditory experiences. Maybe we construct an artificial super-intelligence equipped with high-speed, high-power artificial eyes that survey the operations of its artificial brain, laid out in a network before it. Alternatively, we could imagine that I acquire eyesight so keen that I can discern a flicker in a film projecting at a million frames per second, and at the same time, the cognitive capabilities to process such additional visual information quickly and effectively. We could imagine that my brain evolves, perhaps through a gradual process in which synthetic circuits replace brain cells, into a network that is laid out in front of me, and whose detailed operations and changes I visually observe, just as I can presently observe the changes that characterize the operation of a camshaft.

Suppose that the network consists of a large number of electrical relays, each of which opens and closes depending on the states (opened or closed) of the relays to which it is directly connected. As I observe the network, I see many of these relay events (openings and closings of electrical switches) take place. In particular, there is a sequence of relay events $r_1, r_2, r_3, r_4 \ldots r_{100}$, occurring at times $t_1, t_2, t_3, t_4 \ldots t_{100}$, respectively. I visually observe each of these events.

Assuming the truth of phenomenological serialism, we may suppose that various events in this sequence constitute various conscious experiences. In particular,

[7] I am not the first person to be struck by this; see Meehl (1991). Although the arguments I am about to give focus on hearing and vision, one could easily give parallel arguments for any pair of sensory modalities.

we may suppose that sequence $\langle r_{23}, r_{27}, r_{31} \rangle$ is one of my visual experiences—one of the experiences that I have while visually observing the operation of the relay-network. (Or, if the serialist does not want to *equate* visual experiences with sequences of relay events, we can say that my visual experience is an instantiation of some visual quale or qualia by the $\langle r_{23}, r_{27}, r_{31} \rangle$ sequence.) Call this visual experience V.

Furthermore, and still assuming the truth of serialism, we can suppose that I have a diachronically simple auditory experience that consists of the relay-sequence $\langle r_{10}, r_{16}, r_{22}, r_{28}, r_{34}, r_{40} \rangle$. (Or, if the serialist does not want to equate auditory experiences with sequences of relay events, we can suppose that I have a diachronically simple auditory experience that consists of the instantiation of one or more auditory qualia by the $\langle r_{10}, r_{16}, r_{22}, r_{28}, r_{34}, r_{40} \rangle$ sequence.) Call this auditory experience A.

In the situation described, I have a visual experience in the $[t_{23}, t_{31}]$ time-interval (the interval of time that begins at t_{23} and ends at t_{31}). I also have an auditory experience in the $[t_{10}, t_{40}]$ interval. However, since A is a diachronically simple auditory experience, we may suppose that I have no auditory experience in any sub-interval of the $[t_{10}, t_{40}]$ interval. In particular, we may suppose that the events that occur in the $[t_{23}, t_{31}]$ interval are not sufficient for the occurrence of any auditory experience.

If we accept the serialist's theory, we have to say that the situation just described is a possible one. But there is something very strange about this situation. It is a situation in which there is an interval of time within which I have both an auditory experience (A) and a visual experience (V), despite the fact that I have no audio-visual experience. Ordinarily, this would not pose any problem: it would just mean that A occurred before or after V. But we can't say this about the situation we are contemplating. A does not occur before V, since no sequence of events that exists before the last event in the sequence that constitutes V contains all of the events required for A. And A does not occur after V, since no sequence of events that exists after the first event in the sequence that constitutes V contains all of the events required for A. It would be different if we were to identify the occurrence of A with the last event in the $\langle r_{10}, r_{16}, r_{22}, r_{28}, r_{34}, r_{40} \rangle$ sequence, but that would not be consistent with the serialist position.

In short, serialism entails that there could be a period of time in which you had both auditory and visual experience without having any audiovisual experience, and without having auditory experience before or after visual experience.

We can go farther. It is consistent with the serialist position to suppose that more than one relay-sequence within the $[t_1, t_{100}]$ interval constitutes a visual experience of mine. For instance, it is consistent with serialism to suppose that each of the following sequences is a visual experience:

$\langle r_9, r_{11}, r_{13}\rangle$ $\langle r_{14}, r_{15}, r_{17}\rangle$ $\langle r_{18}, r_{20}, r_{24}\rangle$
$\langle r_{25}, r_{26}, r_{29}\rangle$ $\langle r_{30}, r_{32}, r_{35}\rangle$ $\langle r_{36}, r_{38}, r_{41}\rangle$

We can imagine that these experiences occur as successive parts of an extended visual stream of consciousness, although we needn't suppose that they occur as parts or phases of any single, complex experience (any more than the experiences that I have when I watch the operation of a camshaft for several minutes all belong to a single, complex experience).

In this situation, there is an interval of time (viz. $[t_9, t_{41}]$) in which I have no audiovisual experience—no experience characterized by both visual and auditory phenomenology—despite the fact that I have an auditory experience (A) within that interval, and despite the fact that the interval is divisible without remainder into sub-intervals, each of which contains some visual experience of mine.

In short, serialism entails that there could be a period of time within which I had auditory experience, and throughout which I had visual experience, but within which I had no audiovisual experience.

The fact that phenomenological serialism has these implications is a problem for phenomenological serialism because it puts the theory into conflict with the intuitive and well-defended idea that conscious experience is, in a certain sense, *unified*. Call this the *phenomenal disunity problem* for serialism.

Tim Bayne and David Chalmers express the intuition that consciousness is in a certain way essentially unified by saying that "[n]ecessarily, any set of conscious states of a subject at a time is unified"; they call this the "unity thesis."[8] In their discussion, Bayne and Chalmers take it for granted that conscious experience is a temporal phenomenon. For a more cautious statement of the unity thesis that does not assume this, we can say:

> Necessarily, if consciousness is in time, then the phenomenology that characterizes the conscious experience of any given conscious subject at any given time is unified.

Call this the "cautious unity thesis."[9]

[8] Bayne and Chalmers (2010), 498.

[9] The cautious unity thesis is actually more cautious than Bayne's and Chalmers' unity thesis in two ways. First, because it doesn't assume that consciousness is in time. Second, because it does not apply only to beings that have more than one experience at a time: maybe the phenomenology that characterizes the conscious experience of a conscious subject at a given time is phenomenology that pertains to a single, phenomenally complex experience.

The cautious unity thesis is extremely plausible. Try to imagine having a visual experience as of palm tree and, at the same time, an auditory experience as of a seagull's cry, without having an experience as of both a palm tree and a seagull's cry. Or try to imagine having an olfactory experience as of coffee and, at the same time, a tactual experience as of washing your hands, without having any experience characterized both by the phenomenal smell of coffee and the phenomenal feel of running water. If you are like me, you will fail when you try to imagine these things. For us, it is inconceivable that conscious experience might be disunified in this way. As Bayne and Chalmers put it:

> Much of the reason for accepting the truth of the unity thesis comes from the fact that its denial seems to be inconceivable and perhaps incoherent . . . It is difficult or impossible to imagine a subject having two phenomenal states simultaneously without there being a conjoint phenomenology for both states . . .[10]

Whenever we try to conceive of two experiences occurring simultaneously without occurring as parts of a single, complex experience, we cannot help but conceive of them as occurring in *different* conscious minds—different subjects of conscious experience. Likewise, whenever we try to conceive of two qualia getting instantiated simultaneously without getting instantiated by a single experience, we cannot do so without conceiving of the qualia as characterizing the experiences of different minds or subjects.

These facts weigh heavily in support of the phenomenal unity thesis. Inconceivability is not an infallible guide to impossibility, but it is a decent guide. If a putative state of affairs is prima facie impossible, defies intelligible description, and resists all efforts to demystify, we are entitled to regard that putative state of affairs with suspicion. At the very least, we ought to suspend judgment on its possibility, which means avoiding commitment to theories known to entail that it is possible. In the present context, that means avoiding commitment to phenomenological serialism.

Now, there are cases that might seem to speak against the unity thesis. But none of these rises to the level of a convincing counter-example. For instance, cases of split brains or split personalities might demonstrate the possibility for a single cognitive agent to have simultaneous experiences that are not phenomenally unified. But in describing these cases, there is no need to suppose that any *conscious subject* has phenomenally disunified experiences. Even if we can count a two-headed man as a single cognitive agent, we have no good reason to count him (them?) as a single subject of conscious experience; the same goes for the extreme

[10] Bayne and Chalmers (2010), 537; 514.

cases of split brains or split personalities that are sometimes erroneously supposed to cast doubt on the phenomenal unity thesis.[11]

I have nothing to add to what has already been said in defense of the unity thesis (or the cautious version of it stated above). The thesis seems to me to be as good a candidate as any for an a priori truth. But if the history of philosophy teaches us anything, it is that any claim to a priority will meet with skepticism from one quarter or another. My goal here has not been so much to persuade anyone of the truth of the unity thesis, as to make it clear to those who are already persuaded of its truth that they should reject phenomenological serialism. They should reject it because it implies, in violation of the cautious unity thesis, that someone could have visual experience while having auditory experience, without having any audiovisual experience.[12]

4.4 Russellian dualism

If we reject serialism, but uphold the view that consciousness exists in spacetime, we have no choice but to accept Russell's thesis, and its implication that a conscious mental life is a series of experiential point-events. Russell sees this implication as something of a puzzle, a sign of how the relativistic nature of spacetime "complicates the relation between perception and physics." Russell does not elaborate, but we can take an educated guess at what he had in mind.

For one thing, Russell's thesis undercuts what has become the dominant theory of the relationship between conscious experience and the contents of spacetime: physicalism. According to physicalism, a conscious experience relates to the contents of spacetime by *being* a content of spacetime—a physical content of it.[13]

In a naive conceptual setting, in which time and its determinations are absolute, the physicalist would naturally identify temporally basic experiences with complexes of simultaneous physical events—say, neuron-scale events occurring simultaneously within a human brain. From here, he might build up to more

[11] See Bayne and Chalmers (2010), 514–19, Bayne (2010), and Dainton (2006), 97–112 for relevant supportive discussion. One thing these authors are careful to point out is that there are other, non-phenomenal kinds of unity that the experiences that a given subject has at a given time can fail to exhibit. But the possibility of a failure of one of those kinds of unity does not imply the possibility of a failure of *phenomenal* unity. Many arguments against phenomenal unity hinge on a failure to distinguish between these various kinds of unity.

[12] Again, while I have focused on examples involving auditory and visual phenomenology, we could make the same point by reference to other combinations of experiences (or qualia), including intra-modal combinations.

[13] Or if not by being a physical content of spacetime, then by supervening on, or being grounded in, a physical content of spacetime. The points made below apply to these fancier forms of physicalism no less than to the plain identity form.

complex experiences, but at the most basic level, one would normally expect a physicalistic experience to consist of (or supervene on, or be logically grounded in) a collection of simultaneous physical events distributed throughout some spatially extended network.

But Russell's thesis tells us that our most basic experiences (the diachronically simple ones) are mere point-events. The naive physicalist picture is therefore untenable.

To get around this problem, a physicalist might identify diachronically simple experiences with individual *physical* point-events (or identify such events as the supervenience- or grounding-bases of diachronically simple experiences). The connection of these events to larger-scale physiological activity would have to be spelled out, but as long as there is some exceptionless correlation between a given kind of physical point-event and a given kind of experience, there is a basis (an enterprising physicalist might argue) for equating the two.

Unfortunately for the enterprising physicalist, there is no evidence of such events. What is distinctive about, for example, a human body, is not that it is made up of distinctive subatomic events, but that the subatomic events that constitute it relate to one another so as to form distinctive physical states and processes extended in spacetime. The enterprising physicalist is simply rash if he posits a so-far undiscovered fundamental physical phenomenon to account for a so-far unobserved physiological anomaly.

Assuming that science has not overlooked any physiological anomalies that demand an explanation in terms of an unanticipated extension to basic physics, we are left with a view of ourselves as quite homogeneous with our surroundings at a fundamental physical level. Any physical point-event that occurs in the body of a sentient being can also occur outside of any such body. So, if we identify any of our experiences with *physical* point-events, we will have to identify them with events that occur, or could occur, in rocks, sandwiches, and other seemingly unconscious things.[14]

Furthermore, it is likely that for every subatomic physical event that occurs in the body of a bat (or any other creature), there occurs an exactly similar subatomic event in my body. Here, the implication of the enterprising physicalist's proposal is that there occur in me the same kinds of experiences as occur in bats, squid, and other creatures with unusual powers of perception.[15]

[14] Likewise, if we say that one of your experiences supervenes on, or is logically grounded in, a physical point-event, we'll have to say that it supervenes on or is logically grounded in a point-event of a kind that also occurs in rocks, sandwiches, etc.

[15] A related problem for the enterprising physicalist is that it is not obvious that there is enough variety among physical point-events to reflect the phenomenal variety of diachronically simple experiences. This is related to the so-called "grain problem"; see especially Lockwood (1993), 275–6.

The appeal of physicalism was always its promise to equate conscious mental states with physical mediators of environmental stimulus and behavioral response. A theory that attributes consciousness to a sandwich cannot possibly deliver on this promise.[16]

However, physicalism is not the only theory of consciousness that thinks of conscious experiences as occurring in physical time and space. Naturalistic-minded dualists also think of consciousness as existing in the same spacetime as physical events. One way to do this is by thinking of conscious experiences as instantiations of non-physical qualia by sequences of physical events; this would be a dualistic version of serialism, and would face the disunity problem just like any other serialist theory. Alternatively, a dualist could construe our experiences, or at least our most basic experiences, as non-physical point-events. This would be a dualist version of pointillism.

According to the latter theory, a conscious mental life is a series of durationless and unextended (or at least spatiotemporally structureless) occurrences. Such a theory is essentially a dualist philosophy of mind that satisfies Russell's Thesis. Call it Russellian dualism.

Russellian dualism entails a somewhat weird spatiotemporal structure for conscious experiences, and conscious selves. It implies that we spend almost all of our lives in the condition of zombies. For even if we assume that a single moment of time has some extent, this cannot plausibly exceed the duration of the Planck moment, which is about 10^{-44} seconds (as measured by a terrestrial clock).[17] Thus, if one were conscious continually for the space of one second, one would have within that second at least 10^{44} successive conscious experiences.

Even if we assume, very generously, that we are capable of having experiences as of events as brief as one one-thousandth of a second, you could not have had 10^{44} successive experiences in the past ten seconds without having had experiences with a total phenomenal duration of 10^{41} seconds—about 10^{34} years.

Of course, phenomenal durations are not straightforwardly additive, like physical durations. From the fact, if it were a fact, that in the past second you had 10^{44}

[16] The same difficulties would seem to attend any proposal to equate conscious experiences with fundamental physical entities, on any reasonable conception of what those entities are. So, for example, it would not seem to do the physicalist any good to maintain that the fundamental physical entities are not point-events, but, as Richard Healey suggests, certain closed curves in spacetime (or the instantiation by such curves of certain properties); see Healey (2007), xvii–xix, 21–6, 104–28.

[17] The qualification, which I henceforth leave implicit, is a nod to relativistic time dilation. A moment cannot have a greater duration than the Planck duration, since there are changes that occur in periods of time scarcely longer than this, and a change requires more than one moment in which to occur (otherwise, something could simultaneously have and lack a feature with respect to which it changed).

successive experiences as of 10^{-3} second events, it would not follow that you had an experience as of a 10^{41} ($= 10^{44} \times 10^{-3}$) seconds event (or event-sequence).

Still, if, over the past second, you had 10^{44} successive experiences, surely you would be able to remember having—if not 10^{44}, then at least *very* many—experiences in the past second. But if you are anything like me, the truth is that you can remember having at most 10^3 (more like 10) successive experiences over the past second. Even if you had a million unremembered experiences in the past second for every experience you can remember, that would mean that you were conscious for no more than 10^{-35} seconds over the past second of your life. For the remainder of the past second—that is, for practically all of it—you were completely unconscious. Or so a Russellian dualist must say.

The upshot is that just as a table is mostly "empty space," so too a conscious mental life is mostly "empty time" on the Russellian view—to such an extent that removing *all* the conscious experience from a typical human life and closing the resulting time-gaps would shorten the life by only about 10^{-26} seconds.[18]

These are striking implications for any theory of consciousness to have. But after all is said and done, it is not clear that they deserve to be counted as more than innocent curiosities. There is nothing obviously incoherent about these implications, or with the Russellian account. At any rate, Russellian dualism looks to be the best hope for those who want to keep consciousness in time.[19]

4.5 Conclusion

The idea that physicalism conflicts with the introspectible unity of consciousness is at least as old as Leibniz, and probably much older. It is an idea that the considerations brought to light in this chapter tend to support.

[18] 1 year = 3.2×10^7 seconds. 80 year lifespan = 2.56×10^9 seconds of life. (10^{-35} seconds of experience per second) $\times$ (2.56×10^9 seconds) = 2.56×10^{-26} seconds of experience lifelong.

In Chapter 5, we'll see that a temporalist about conscious experience must say that we have, in addition to diachronically simple experiences, diachronically complex experiences consisting of temporal sequences of diachronically simple experiences, and that a stream of consciousness consists of a series of temporally overlapping diachronically complex experiences. This commits the temporalist to the view that—not just a conscious mental life, but—many of our actual conscious experiences are mostly empty time. While not necessarily phenomenologically unrealistic—an absence of experience is not to be confused with an experience of absence—this does make the temporalist picture of consciousness even more peculiar.

[19] This is not to say that Russellian dualism is problem-free: it is not. But it is not clear that the problems facing Russellian dualism are much more serious than those facing more traditional forms of dualism. For a critical discussion of Russellian dualism, see Chalmers's remarks on "identity panpsychism" in Chalmers (2012).

At the same time, we have found that even a dualist has to make some tough choices if he is to respect the unity of consciousness without abandoning the common sense notion that conscious experiences occur in the same time as physical events. In Chapter 5, we shall see dualism constrained even further by facts relating to the stream of consciousness.

5

The stream of consciousness

> But all my hopes vanish, when I come to explain the principles, that unite our successive perceptions in our thought or consciousness.
>
> David Hume, *Appendix to A Treatise of Human Nature*

5.1 Introduction

A stream of consciousness is a series of experiences whose adjacent members relate to one another as my present state of mind relates to the state of mind I was just in, rather than as my first mental state this morning related to my last mental state last night, or as my present state of mind relates to the state of mind that *you* were just in. The exact nature of this relationship remains to be seen; here, I am not offering an analysis of the concept of a stream of consciousness, but merely drawing attention to the phenomenon that I wish to discuss.[1]

I wake up, eat breakfast, brush my teeth, take a shower, get dressed, walk to campus, check my e-mail, walk to the lecture theater, give a lecture, drink another cup of coffee, and check my e-mail again. I am awake the whole time, and the experiences I have at each stage of my morning belong, we may suppose, to a single stream of conscious experience. But there is, for me, no such thing as an *experience* as of waking up, eating breakfast, brushing my teeth, taking a shower, getting dressed, walking to campus, checking my e-mail, walking to the lecture theater, giving a lecture, drinking a cup of coffee, and then checking my e-mail again. I have no experience characterized by both shower-taking and lecture-giving phenomenology. The shower-taking qualia and the lecture-giving qualia all contribute to a single stream of consciousness, but not by contributing to a single experience—a single conscious state of mind.

[1] When I speak of "my present state of mind," I mean the state of mind that I am attempting to refer to by my utterance of the words "my present state of mind," making no presumption as to the temporality or otherwise of the mental state in question. Likewise, I make no assumption of temporality when I speak of "the sensation I just had," "my last conscious experience last night," etc.

Now, the word "experience" has several meanings, and it might be that in one of them, the statement that all my day's sensations belong to a single experience comes out true; presumably Michael Tye is using the word with some such meaning when he claims that all of the experiences in a stream of consciousness belong to a single, temporally extended experience. But it should be uncontroversial that the visual qualia that characterize the experience I have when I glance at an apple "belong to a single experience" in a sense in which the various qualia that contribute to an extended stream of consciousness (during which I watch the apple being washed, cored, baked, and eaten) do not. It is this former sense of "experience" that is relevant to our present purposes.[2]

How do the experiences that make up a stream of consciousness relate to one another? By virtue of what do different experiences belong to the same stream of consciousness? To answer this question, it helps to compare cases in which there is, intuitively, a stream of consciousness with similar cases in which there intuitively is not.

I am sitting on a kitchen stool, struggling to open a jar of pickles, when suddenly I find myself—or so it seems to me—seated on a mossy log in the middle of a primeval forest, holding a large pebble in one hand and a partially-knapped spearhead in the other. The cold fluorescent glare of the kitchen transforms into soft sunlight filtered through a leafy canopy high above me; the construction noise from the street outside gives way to the sound of a woodpecker echoing through the trees. Underfoot I feel the hard kitchen tile replaced by springy earth; as I inhale, the smell of baked beans yields to the tannic scent of moldering humus. I am understandably shocked and bewildered.

Compare this with a scenario in which my kitchen experiences are abruptly replaced by experiences phenomenally indistinguishable from those of a certain Native American from a thousand years ago: the same bodily sensations, the same perceptual phenomenology, the same cognitive phenomenology, etc. In this case, there is no shock or bewilderment, since these feelings characterize neither my kitchen experiences nor the pre-Columbian experiences, which are the only experiences there are in this case.

[2] Dainton makes this point in Dainton (2008), 71–3. For Tye's comments, see Tye (2003), 108. A separate question is whether we have complex *experiences* consisting of distinct, simpler experiences. One might deny (as Tye does) that we have complex experiences without affirming (as Tye also does) that a stream of consciousness is a single, extended experience. In particular, one might hold that every conscious experience is diachronically simple (in the sense defined in Chapter 3) without counting an entire stream of consciousness as a conscious experience. Instead, you could say that a stream of consciousness is a sequence of distinct diachronically simple experiences that do not collectively constitute any further experience.

In the first scenario, kitchen phenomenology streams into forest phenomenology. In the second, it does not; rather, it is as if I had been annihilated and someone else had been created in my place. The explanation of this difference is that in the first case, but not the second, I have experiences as of changes from kitchen to forest: changes in lighting, sound, smell, posture, etc. The presence of these experiences in the first scenario is what binds the earlier kitchen experiences with the later forest experiences into a single stream of consciousness.

Let us consider how this works in detail. In the first case, I have: (1) an experience as of sitting in the kitchen then still sitting in the kitchen; (2) an experience as of sitting in the kitchen then sitting in the forest; and (3) an experience as of sitting in the forest then still sitting in the forest. In the second case, I have (1) and (3) but not (2). There is what we may call *phenomenal overlap* between (1) and (2), and between (2) and (3), but not between (1) and (3): the phenomenal content of experience (2) is, so to speak, partially offset toward the future relative to the phenomenal content of experience (1), and likewise for (3) and (2).

In general, two experiences phenomenally overlap if, and only if, one of the experiences is completely describable (in phenomenal terms) as an experience as of *X* then *Y*, and the other completely describable as an experience as of *Y* then *Z*, for some *X*, *Y*, and *Z*. The conscious experiences I have in the first case described above belong to a series of experiences, each of which (save the first) phenomenally overlaps the preceding experience in the series. Let us call a series of experiences that satisfies this condition a *phenomenally integrated* series of experiences. The experiences I have in the first case form a phenomenally integrated series; the experiences I have in the second case do not. That is why the experiences in the second case do not all belong to a single stream of consciousness.

Any series of conscious experiences all of whose adjacent members phenomenally overlap is a phenomenally integrated series. The experiences that I have in the first scenario described above, but not those that I have in the second, form a phenomenally integrated series. In both scenarios, I begin with an experience as of struggling with a jar of pickles in a modern kitchen. But in the second scenario, this experience is followed—not by an experience as of the scene changing from kitchen to forest, but—by an experience that exhibits no kitchen-to-forest phenomenology at all. That is why it is, phenomenologically, as if someone sitting in a kitchen had been annihilated and someone sitting in a forest had been created in his place.

The two scenarios we are discussing differ in point of phenomenal overlap, but that is not the only respect in which they differ. Another difference is that in the first scenario, but not the second, the forest experiences come with a conscious memory (or apparent memory) of having just been sitting in a kitchen struggling with a jar of pickles. Might this, rather than phenomenal overlap (or its absence) be the real reason why the experiences in the first scenario, but not those in the second, belong to the same stream of consciousness?

Suppose that the state of mind that replaces my kitchen experiences in the second case is a state of mind phenomenally indistinguishable from some actual past state of a pre-Columbian Native American, with the exception that it includes a conscious memory (or apparent memory) of having just been sitting in a kitchen grasping a jar of pickles. The existence of this memory might give us a reason to say that the forest experiences belong to the same *person* or conscious self as the kitchen experiences, but it does not give us a reason to say that the forest and kitchen experiences belong to the same *stream of consciousness*. If there is no phenomenal overlap between the two experiences (modern kitchen, primeval forest), then we still have a break in the stream of consciousness: it would be as though I had suddenly fallen asleep or passed out while struggling with the pickle jar, and then suddenly awoken in the middle of a forest grasping a stone, without ever having any experience as of a change of scene. Two experiences need not belong to a phenomenally integrated series in order to belong to the same subject, but they must belong to a phenomenally integrated series in order to belong to the same stream of consciousness.[3]

Phenomenal integration is necessary, but not sufficient, for phenomenological streaming. If I have a stream of consciousness consisting of successive experiences $I_1, I_2, I_3, \ldots I_n$, and my counterpart on Twin Earth has a phenomenologically indistinguishable stream consisting of experiences $T_1, T_2, T_3, \ldots T_n$, then the series $I_1, T_2, I_3, T_4, \ldots \mathrm{I}_{n-1}, T_n$ is phenomenally integrated, but it isn't a stream of consciousness. Later, we'll consider what we have to add to phenomenal integration to get a set of conditions not just necessary, but sufficient for a stream of consciousness. Our immediate concern, however, is a problem that arises from the fact that phenomenal integration is necessary for the stream of consciousness.

5.2 The paradox of the stream of consciousness

Suppose that I am listening to someone play scales on the piano, and suppose (somewhat unrealistically) that the only conscious experiences I have while listening are auditory experiences of the various notes being struck. We might try to represent my stream of consciousness in this situation as follows:

[3] It is also unclear that a human stream of consciousness must contain any conscious memories, unless we count the existence of certain instances of phenomenally overlapping cognitive experience as sufficient for conscious memory, and hold that all human experience is characterized by cognitive phenomenology. See Dainton (2006), 123–7 for relevant discussion.

(1) *Do, Re, Mi, Fa, So, La, Ti*

where "*Do*" designates an experience as of the note C being struck, "*Re*" an experience as of D being struck, etc. This representation is inadequate, however, because it does not capture the phenomenal integration of the stream of experiences it attempts to represent. We might equally well use (1) to represent the series of experiences I would have if I were first in a mental state indistinguishable from that of someone who hears only a solitary C, and then in a mental state indistinguishable from that of someone who hears only a solitary D, and then in a mental state indistinguishable from someone who hears only a solitary E, etc.[4]

A seemingly natural fix is to represent my auditory stream of consciousness like this instead:

(2) *Do . . . Re, Mi . . . Fa, So . . . La, Ti*

where "*Do . . . Re*" represents an experience as of C followed by D, "*Mi . . . Fa*" an experience as of E followed by F, etc. But this will not do either. For while it is true that my stream included an experience as of C followed by D as well as an experience as of E followed by F, it *also* included an experience as of D followed by E; but this latter experience goes unrepresented in (2).

Well, let's just insert a new element into the series, to represent the D-followed-by-E experience (as well as the other, similar experiences that (2) omits):

(3) *Do . . . Re, Re . . . Mi, Mi . . . Fa, Fa . . . So, So . . . La, La . . . Ti*

But what we have now represented is not the stream of conscious experience that I had while listening to the scales, but the stream that I would have had if I had heard someone play the following sequence of notes: CDDEEFFGGAAB.

In fact, (3) doesn't even succeed in representing *that* counterfactual stream! This is for the same reason that (1) fails to represent my actual stream. To get an adequate representation of the stutter-scale stream, we could try:

(4) *Do . . . Re, Re . . . Re, Re . . . Mi, Mi . . . Mi, Mi . . . Fa, Fa . . . Fa, Fa . . . So, So . . . So, So . . . La, La . . . La, La . . . Ti*

But it's clear where this is going. In order for the experiences I have while hearing the scales to belong to a single a stream of consciousness, they must constitute a phenomenally integrated series. But in order to constitute a phenomenally integrated series, it seems my experiences must include many that I simply do not have—and even if we suppose that I *did* have such experiences, we *still*

[4] This point is due to Phillips (2010), 180. Phillips attributes the point to Husserl.

wouldn't have a stream of consciousness without positing yet further experiences, ad infinitum.

Of course, it would be different if, while listening to the scales, I had just a single, lengthy experience: an experience as of C followed by D followed by E followed by F followed by G followed by A followed by B. But I had no such experience. (Or if I did, we need only consider a longer stretch of time over which I listen to the player play the scales repeatedly.)

And so we have a paradox. In order for our experiences to occur as parts of streams of consciousness, our experiences must be phenomenally integrated, without collectively constituting a single experience. But when we try to describe a stream of consciousness in a way that respects these constraints, we end up badly misdescribing it as including many experiences that it does not, in fact, include.

All of the most obvious ways of trying to solve the paradox of the stream of consciousness really only succeed in masking it; let us consider some of these unsuccessful attempts now.[5]

5.2.1 Divided appearances

One might try to solve the paradox by suggesting that my first experience is an experience as of a whole *Do* followed by a half *Re,* my second an experience as of a half *Re* followed by a half *Mi,* my third an experience as of a half *Mi* followed by a half *Fa,* etc. The idea here is that by halving the phenomenal tones, we can accept the implication of repeated contents without misrepresenting the stream of consciousness:

(5) *Do*... *Re*/2, *Re*/2... *Mi*/2, *Mi*/2... *Fa*/2, *Fa*/2... *So*/2, *So*/2... *La*/2, *La*/2... *Ti*/2, *Ti*/2

However, it could be that I had no experience as of a half-note followed by a half-note. It could be that while listening to the piano, I hear (and seem to myself to hear) only whole notes occurring in succession. This is particularly clear in the case where my experience is as of very brief tones. As we found in Chapter 3, there is a lower bound on how brief a tone (or anything else) can appear to me. If the scales are played quickly enough, I may have an experience as of *Do* followed by *Re,* where I cannot have any experience as of a tone briefer than this *Re.* In this case, we cannot say that my stream of consciousness included an experience as of *Do* followed by a half-*Re,* a half-*Re* followed by a half-*Mi,* etc.

Furthermore, and regardless of the phenomenal durations of the experiences I have while listening to the piano, I *do* have an experience as of a whole *Do*, as well

[5] Dainton criticizes some of these proposed solutions to the paradox in Dainton (2006), 123–7, 141–2.

as an experience as of a whole *Re*, etc. But this is not represented in (5), which could depict the mental life of someone who never had any experience as of a whole *Re*; for example, we could use (5) to represent the experiences of someone who has an experience phenomenally indistinguishable from that of someone who hears only a C followed by a half-D, and then an experience phenomenally indistinguishable from someone who hears only a half-D followed by a half-E, etc.

5.2.2 The memory theory

We have already considered, and rejected, the suggestion that memory rather than phenomenal overlap does the work of holding successive experiences together in a single stream of consciousness. But one might suggest that phenomenal overlap itself could be understood in terms of memory, or more exactly, a correspondence between the sensory qualia of one experience and the memorial qualia of a succeeding experience. According to this proposal, my *Re . . . Mi* experience does not instantiate the same *Re*-quale as my *Do . . . Re* experience, but only a retrospective or (as I shall say) "memorial" relative of it. We might describe the *Re . . . Mi* experience as being, among other things, an experience as of the D tone having just recently sounded. The memorial *Re*-quale need not even be a strictly auditory quale. Think of how the sense of silence that follows the end of a symphony differs from the sense of silence that comes with standing motionless in the middle of a frozen field on a still winter night. Arguably, these two perceptual silences differ by virtue of the fact that the first, but not the second, involves the instantiation of auditory memorial qualia.[6]

But even if my experience while listening to the scales is partly characterized by memorial qualia, that doesn't change the fact that it is *also* characterized by an experience as of a bold and clearly struck C followed by a bold and clearly struck D, as well as an experience as of a bold and clearly struck D followed by an equally bold and clearly struck E. The phenomenology corresponding to each tonal change is as much—and as vivid—a part of my auditory experience as the phenomenology corresponding to each tone's loudness or pitch. A more nuanced description of my experience might also bring in memorial qualia, but these can only augment, and not supplant, the more salient features of the experience.

[6] The memory theory is a version of what Dainton calls "retentionalism"; see Dainton (2006), 150–61. There is a nice critical discussion of memory theories (recent and vintage) in Hoerl (2009), 2–5.

5.2.3 *The inner-awareness theory*

As an alternative to the memory theory, we might try to solve the paradox—not by replacing the experiences depicted in (1) with other experiences, but—by supplementing them with additional experiences.

According to this "inner-awareness theory," when I listen to the scales, I have a single, simple experience for each note that gets played. Depiction (1) is therefore correct, as far as it goes. However, in addition to these simple experiences, I have a number of further, complex experiences. I have these further experiences by being consciously aware of successive overlapping sequences of my simple experiences.

Figure 5.1 represents this supposed state of affairs by placing successive overlapping sequences of simple experiences within the scope of successive acts of inner-awareness. Each inverted angle represents an act of inner-awareness of a sequence of simple experiences, where each act occurs at the same time as the last simple experience that falls within its scope. For example, A_1 is an introspective apprehension of experiences *Do* and *Re*, occurring at the same time as *Re*; A_2 is an introspection of *Re* and *Mi*, occurring at the same time as *Mi*; and so forth.[7]

What are these acts of inner awareness? Presumably they are involuntary, since our experiences occur in streams independently of our will. They must also be conscious acts: if they contributed nothing of their own to the phenomenology of my experience, they could not account for the fact that I have an appearance of C followed by D, rather than just an appearance of C followed by an appearance of D.

One problem with the inner-awareness theory is that it is far from clear that one must be aware of one's experiences in order to have a stream of consciousness. Normally, when I go sailing, I pay no attention to my conscious experiences: I'm preoccupied by the wind, the current, and the set of the sail. Nevertheless, I do

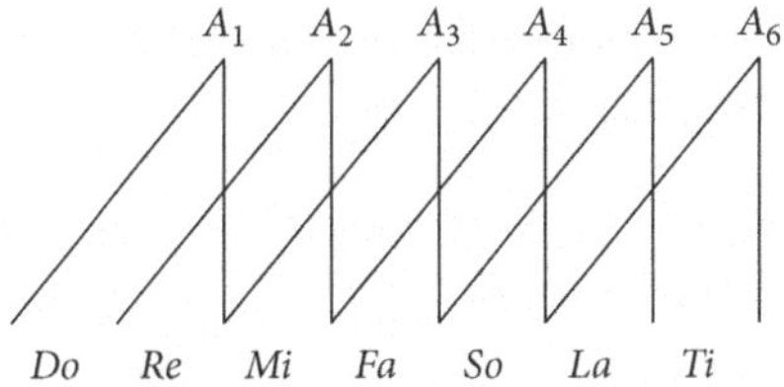

Figure 5.1 Inner-awareness theory.

[7] See Broad (1923), 344–58, and the discussion in Dainton (2006), 136–61.

have experiences, and it would seem that these experiences belong to my stream of consciousness.

Furthermore, acts of introspection or states of inner-awareness seem to be the sorts of things that can themselves belong to a stream of conscious experience. But it is hard to see how the inner-awareness theory can account for their presence in a stream without introducing a higher level of introspective states, which would in turn seem equally capable of joining the stream. The inner-awareness theory therefore faces the prospect of a vicious regress.

These objections are not necessarily fatal. A partisan of the theory might simply deny that one has a stream of consciousness—or even that one is conscious at all—when unaware that experiences are taking place. To cut off the regress, he might deny that conscious acts of inner awareness occur in a stream of conscious experience, holding that only lower-level sensations occur in a streaming fashion.

But the main problem with the inner-awareness theory is, again, the problem of repeated contents. Since the posited acts of awareness are themselves experiential entities, we can ask: what are their phenomenal properties? Since A_1 is an introspection of *Do* and *Re*, its phenomenal quality is partly characterized by the auditory phenomenality of *Re*. But A_2 is also an introspection of *Re* (as well as *Mi*). So it looks as though according to the inner-awareness theory, I hear *Re* played twice, whereas in fact I hear it just once.

One might suggest that A_2 is an introspection of *Re* as past, while A_1 is an introspection of *Re* as present. This is like the memory theorist's appeal to memorial qualia, and the response is the same: in order for these acts of inner-awareness to facilitate phenomenal integration, they must include an experience as of a bold and clear D following a bold and clear C, as well as an experience as of a bold and clear E following a bold and clear D. Since A_1 occurs before A_2, the *Do . . . Re* experience that A_1 supposedly constitutes must occur before the *Re . . . Mi* experience that A_2 supposedly constitutes. The implication is that a bold and clear *Re*-quale gets instantiated twice in succession while I listen to the scales, whereas in fact it gets instantiated only once.

5.3 A temporal solution

So far, we have considered a variety of failed attempts to solve the paradox of the stream of consciousness. We now consider a better attempt, due to Barry Dainton.

Dainton proposes to solve the paradox by holding that the pairwise phenomenally overlapping experiences that make up a stream of consciousness also overlap in the literal sense that they have parts in common. For example, according to

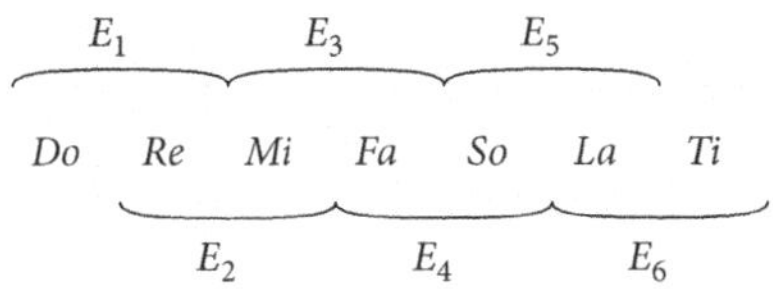

Figure 5.2 Temporal overlap theory.

Dainton, my *Do . . . Re* experience has two briefer experiences as parts: (i) an auditory experience as of C, and (ii) a slightly later auditory experience as of D. Similarly, my *Re . . . Mi* experience comprises: (iii) an experience as of D, and (iv) a slightly later experience as of E. But—and this is the key idea behind Dainton's theory—experience (ii) is one and the same experience as experience (iii). So the *Do . . . Re* experience and the *Re . . . Mi* experience have between them only three basic constituents: an auditory "Do" experience, an auditory "Re" experience, and an auditory "Mi" experience. Since I do not, by this account, have two experiences as of "Re," the account escapes the problem of repeated contents.[8]

Dainton calls this the "overlap model" of the stream of consciousness. In effect, the theory construes phenomenal overlap as a species of objective temporal overlap. According to the overlap model, a stream of consciousness is made up of temporally extended experiences, where each experience in the stream has some (but not all) of its temporal stages in common with the experiences that precede or follow it.[9]

For example, according to the overlap theory, we can depict my stream of consciousness while listening to the piano with the diagram in Figure 5.2. In this diagram, the presence of a brace over or under two successive experiences indicates that these experiences are "co-conscious," forming a further, complex experience that has each of the two experiences as a part. I have chosen to depict these complex experiences—E_1, E_2, E_3, etc.—as comprising two simple experiences each, but this is merely for ease of exposition. I am just assuming, for simplicity's sake, that as I listen to the scales, I have an experience as of C followed by D, and no experience as of C followed by D followed by E. If I do have an experience as of C followed by D followed by E, an experience as of D followed by E followed by F, etc., the overlap theory gives the following picture:

[8] Dainton (2006), 162–77.

[9] See Dainton (2006), 162–82. Dainton attributes the basic idea behind the overlap model to John Foster: see Foster (1979), Foster (1982), 259–60, and Foster (1991), 245–50.

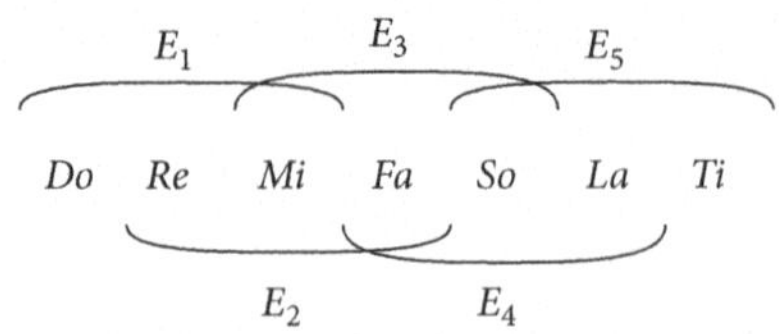

Figure 5.3 Longer overlapping experiences.

It might also happen that different diachronically complex experiences in a stream of consciousness comprise different numbers of successive simple experiences; this might happen if, for example, the pianist plays the scale more quickly as he goes along.

Actually, we could have recycled Figure 5.1 to illustrate Dainton's theory, by interpreting A_1, A_2, etc. as experiences comprising the simple experiences they respectively subtend, rather than as conscious acts of awareness of those experiences. The key difference between the overlap theory and the inner-awareness theory is that according to the former, but not the latter, successive appearances of auditory transition, such as *Do . . . Re* and *Re . . . Mi,* have one of their constituent simple experiences in common. According to the inner-awareness theory, the *Re*-quale gets instantiated at two different times as I listen to the scales. According to the overlap theory, it gets instantiated only once. Hence, no repeated contents.

Dainton's solution to the paradox of the stream of consciousness relies on the idea that some of our experiences are extended in time, consisting of successive, briefer, and (ultimately) diachronically simple experiences, none of which itself has successive experiences as parts. In Dainton's terms, a stream of consciousness consists of overlapping sequences of "diachronically co-conscious" experiences. (Two experiences are diachronically co-conscious just in case they are non-simultaneous parts or phases of a temporally extended experience.)

Dainton does not consider diachronically complex experiences necessary to account for the existence of subjective appearances of succession and change. He acknowledges that even our temporally basic experiences typically exhibit phenomenal change, despite not consisting of temporal sequences of experiences. The overlap theory does not deny that we have diachronically simple appearances of succession and change, but it does deny that *all* of our appearances of succession and change are diachronically simple.

Whether or not individual diachronically simple experiences have any duration, a diachronically complex experience persists from the first to the last of its constituent simple experiences. It is tempting to think that this objective duration must somehow bear on the subjective, *phenomenal* duration of the complex experience. This, however, is a temptation that a proponent of the overlap theory should resist.

The objective duration of a diachronically complex experience, like that of any other sequence of events, is not absolute. Described in terms of one frame of reference, a given sequence of experiences might occur over a period of half a second; described in terms of another, the same sequence might occur over a period of minutes, days, or years. Both descriptions may be equally true to the underlying facts; both may belong to fully accurate representations of reality.

By contrast, the *subjective* duration of an experience is entirely absolute. If I have an experience as of three flashes occurring in rapid succession, this is a fact that any complete and accurate description of reality must register. If someone says that my experience was as of three flashes occurring several minutes apart from one another, he badly misdescribes my experience.

If we were to say that my experience depended for its phenomenal character on the occurrence in rapid objective succession of three experiences—three visual images of flashes—we would be demoting the phenomenal character of the experience to a mere artifact of representation. The phenomenal duration of an experience is absolute; its objective duration, if any, is not. Therefore, we cannot say that the objective duration of a diachronically complex experience has any bearing on its subjective duration. The objective duration of an experience is phenomenologically meaningless.[10]

None of this constitutes a problem for the overlap model: it is just further evidence of the looseness of the connection between the subjective quality of an experience and the experience's supposed objective spatiotemporal structure. However, the fact that the overlap model relies on diachronically complex experiences has one implication that should give us pause: it entails a violation of an otherwise universal symmetry of nature. Let me explain.[11]

Our mental lives contain a huge variety of conscious experiences. But within this teeming variety there is considerable order. By and large, our experiences occur in a way that permits us to interpret a great many of them as veridical perceptions of a universe of events unfolding in conformity with certain laws—the laws of physics.

The possibility of interpreting our experiences this way is more obvious for some laws than others. We can hardly help interpreting our experiences as revealing to us a world of spatiotemporal continuants. If we think of the existence of such

[10] On this point, at least, I agree with Arnold Zuboff: see Zuboff (1977–8), 143.

[11] For an earlier version of the argument that follows, see Pelczar (2014).

continuants as a law of nature, we can say that our experiences are obviously interpretable as including perceptions of a world that conforms to this law.

Our experiences are also interpretable as including veridical perceptions of a world that conforms to the Einstein Field Equations. The possibility of interpreting them this way is far less obvious; otherwise, it wouldn't have taken us so long to interpret them thus.

Between these extremes, there is the possibility of interpreting our experiences as revealing a world in which an object at rest tends to remain at rest and an object in motion tends to remain in motion. And there are other intermediate cases as well, some closer to the law of continuants, some closer to the Field Equations.

Not all of our experiences are interpretable as perceptions of things conforming to the laws of physics. Bodily sensations, like tickles and pains, are not. In dreams, we may have sensory or quasi-sensory experiences that resist interpretation as perceptions of any lawlike state of affairs. Not even all of our *perceptions* are interpretable as veridical experiences of law-abiding things. A professional illusionist might make his audience misperceive perfectly law-abiding objects as objects that defy the law of gravity. Still, taken together as a whole, the totality of all human experiences does bear interpretation as including myriad veridical perceptions of states of affairs that conform to various laws of physics.

We can think of the amenability of human experience to such interpretation as a law in its own right, a "law of experience." Or we can speak of different laws of experience, corresponding to different interpretive possibilities. For example, we might speak of a "phenomenological law of gravitation." Our experiences conform to this law by tending to occur in such a way as to admit of interpretation as including veridical perceptions of a physical universe that conforms to the (physical) law of gravitation. Likewise, we can speak of a phenomenological law of the conservation of energy—a law to which our experiences conform by tending to occur in ways that make them interpretable as including veridical experiences of a physical universe in which energy is conserved.

The laws of experience are undoubtedly laws of some kind. But are they fundamental laws?

They are almost certainly not fundamental laws, if physicalism is true. Presumably, if physicalism is true, the laws of experience govern our conscious mental lives due to the fact that some more basic physical laws govern our physical natures.

But we have seen that it is difficult to locate consciousness in spacetime in a way that is both plausible and compatible with physicalism. So let's ask whether the laws of experience are fundamental in a non-physicalistic context. There are two options to consider: dualism and idealism.

As we shall see in subsequent chapters, idealism takes phenomenological laws like the laws of experience to be not just fundamental, but architectonically central. At least, this is true of any defensible idealism.

What about dualism? Obviously, a dualist cannot say that the phenomenological regularities corresponding to the laws of experience reduce to more basic physical regularities. If, for example, the phenomenological law of least action is non-fundamental in a dualist context, it can only be by reducing to some more basic *phenomenological* law (or laws).

What might these more basic laws be? Unlike what I have called the laws of experience (the phenomenological law of gravitation, the phenomenological laws of electrodynamics, etc.), they would have to be laws that we could think of our experiences as conforming to, other than by thinking of our experiences as amenable to interpretation as veridical experiences of a physical universe obeying various physical laws. They would have to be something like "laws of raw feels," corresponding to patterns in our experience that we could describe in low-level phenomenal terms.

Is human experience governed by such laws? While it does not seem that we can completely rule this out, it is a possibility that seems far-fetched, to say the least. To suppose that such laws govern our experience would be like supposing that we could describe a lithographic reproduction of a Turner landscape in terms of low-level laws for the positioning of colored ink-dots—laws that we could formulate and apply without reference to streams, forests, mountains, etc., or depictions thereof. (Such a law might take the form of an algorithm for computing the hue of any given dot from the hues of adjacent dots.) But if there were such low-level regularities in our experience, surely we would have discovered at least some of them by now; we would have formulated at least the rudiments of an algebra of experience that, in its complete form, would encode all information about human experience, including the fact that our experience lends itself to interpretation as containing veridical perceptions of a physical world that obeys the various laws of physics.

If we want to keep consciousness in time on terms that respect the unity of consciousness, we should probably abandon physicalism, and if we abandon physicalism, we should probably regard the laws of experience as fundamental laws (although not necessarily as the *only* fundamental laws, or even as particularly important laws).

Now let's return to the fundamental laws of *physics*. These all exhibit a certain symmetry. In classical Newtonian mechanics, the fundamental laws of motion are time-reversal invariant, in the sense that they would also hold good in a time-reversed counterpart of our world; that is, a world in which events occurred in the opposite temporal order from that in which they actually occur, and with opposite time-dependent properties from those which they actually possess (e.g. opposite velocities).

The situation is different, but similar, in contemporary physics. It turns out that there are physical processes governed by physically fundamental laws that are *not* time-reversal invariant. A time-reversed counterpart of our universe would violate certain laws that determine the probabilities for various forms of elementary particle decay.[12]

Still, even these laws are invariant under a more general reversal of time, spatial position, and charge. So even though a merely time-reversed counterpart of our universe would violate certain fundamental laws of physics, a counterpart of our universe that was reversed with respect to time (as described above), space (as in a mirror reflection), and charge (with positive and negative charges interchanged) *would* obey the same fundamental physical laws as the actual world.

Insofar as our universe is governed by the known laws of physics, it is symmetrical in this way; the fundamental laws of physics are, as they say, "CPT-reversal invariant."

We can think of CPT-reversal invariance as a second-order law: a law to the effect that all first-order fundamental laws of nature are CPT-reversal invariant. The fundamental laws of physics all conform to this second-order law. How about the laws of experience?[13]

If, as Dainton's overlap theory of the stream of consciousness implies, our streams of consciousness consist of sequences of diachronically co-conscious experiences, then the laws of experience are *not* CPT-reversal invariant.

Let us call the CPT-reversed counterpart of the actual world "Unworld." Unworld contains all the same diachronically simple experiences as the actual world, assuming (with Russell) that each of these experiences is a mere point-event. If there actually occurs a diachronically simple experience as of a bird gliding a short distance from left to right, then the same experience occurs in Unworld.

The situation might be different if diachronically simple experiences were, as the serialist maintains, temporally extended sequences of non-experiential events;

[12] The experiments that originally established the violation of time-reversal invariance are described in Lee et al. (1956) and Christenson (1964).

[13] Not *all* regularities commonly described as natural laws are CPT-reversal invariant; for example, the Second Law of Thermodynamics is not. But the Second Law of Thermodynamics is not a *fundamental* law of nature: it describes a regularity that owes its existence partly to the peculiar boundary conditions of our universe (in particular, to the fact that matter was distributed extremely uniformly around the time of the Big Bang). See Price (1996) and Callender (2004) for illuminating discussion.

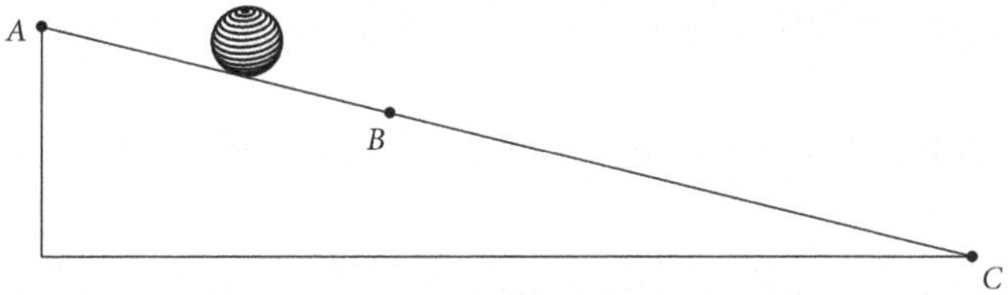

Figure 5.4 Inclined plane experiment.

in that case, one could argue that a reversal in the temporal order of those non-experiential events would involve a corresponding reversal of the phenomenology of the experience that those events constituted. But if, as a Russellian dualist maintains, diachronically simple experiences are confined to individual points of spacetime, they have no spatiotemporal structure to reverse.

Now, in Unworld, our diachronically simple experiences occur in the opposite temporal order from that in which they actually occur in us. Suppose I am watching a ball roll down an inclined plane. And suppose that while watching, I have, among others, a diachronically *complex* experience as of the ball rolling from A to B to C. This experience, we may assume, consists of two diachronically simple experiences: (1) an experience as of the ball rolling from A to B, and (2) an experience as of the ball rolling from B to C (see Fig. 5.4). So, in Dainton's terms, (1) and (2) are diachronically co-conscious experiences that constitute my experience as of the ball rolling from the top to the bottom of the inclined plane.

In Unworld, the temporal order of these two experiences is reversed. Therefore, if I have a diachronically complex experience consisting of them, it is an experience as of the ball rolling first from B to C, then from A to B.

Such an experience need not correspond to a physically impossible occurrence. One could imagine that the ball's constituent particles decohere and recohere in astronomically unlikely, but physically possible, ways, or that the inclined plane set-up is part of some kind of clever smoke-and-mirrors magic trick. But in Unworld, *all* diachronically complex experiences—which means a vast number of the Unworlders' experiences—are like the bizarre ball-rolling experience just described. If, in the actual world, I have a diachronically complex experience as of two initially adjacent magnets flying apart from one another to a distance of ten centimeters, my counterpart in Unworld has an experience as of the magnets going from (say) five centimeters apart to ten centimeters apart and then from zero centimeters apart to five centimeters apart. And so forth.

Considered as a whole, the Unworlders' experiences do not admit of interpretation as veridical perceptions of a world of things conforming to the laws of physics. As the foregoing examples illustrate, this includes even the most basic laws, such as those by virtue of conformity to which physical objects enjoy spatiotemporally continuous existences.

Is this too hasty? Could one argue that, experientially, the difference between our world and Unworld is like the difference between our world and a world in which everyone's visual experience is phenomenally color-inverted with respect to ours? In an inverted-spectrum world, people's experiences do, arguably, conform to the laws of experience, despite differing from our experiences in some of their phenomenal properties. But that is because the experiences in an inverted-spectrum world are interpretable as veridical perceptions of a world that obeys the same physical laws as the actual world. The mental lives of people who are phenomenally color-inverted relative to us needn't contain any appearances of physically impossible events; they are like maps that depict land in blue and water in green, rather than vice versa. By contrast, the mental lives of our counterparts in Unworld *do* contain appearances of physically impossible events; they are like maps that depict oceans having the topography of mountain ranges.

The most we might be able to say for the Unworlders' experiences (assuming they are as described above) is that they bear interpretation as including myriad *misperceptions* of things conforming to the actual laws of physics. There is (so to speak) a permutation of the Unworlders' experiences—viz. the totality of actual human experiences—that does obey the laws of experience. That doesn't change the fact that the Unworlders' experiences themselves do not obey those laws, if the Unworlders' experiences are as the overlap theory predicts.

Of course, Unworld does obey the laws of physics. In Unworld, balls, magnets, and all other objects conform to fundamental physical laws. But if, as Dainton's theory of the stream of consciousness requires, our streams of consciousness consist of overlapping diachronically complex experiences, Unworld does *not* conform to the laws of experience. A proponent of Dainton's theory must therefore regard the laws of experience as fundamental laws of nature that, unlike other fundamental laws, violate CPT-reversal invariance.

In order to keep the stream of consciousness in time, we have to posit diachronically complex experiences. But in order to posit diachronically complex experiences, we have to accept the violation of an otherwise universal symmetry of nature. The upshot is that we cannot think of consciousness as a temporal phenomenon without thinking of it as extending through time differently from all other temporal phenomena.

One reaction to this would be simply to conclude that CPT-reversal invariance is not sine qua non for fundamental lawhood. After all, people used to think that all fundamental laws were time-reversal invariant. Taking consciousness seriously means being open to the idea that phenomenological discoveries might have far-reaching ramifications for the fundamental structure of the world; maybe the failure of CPT-reversal invariance is one such ramification.

At the same time, any dualist of a naturalistic bent will surely have misgivings about a theory that turns consciousness into the unique breaker of what is arguably the most basic and generic of all natural symmetries. If the cost of keeping the stream of consciousness in spacetime is acceptance of a profoundly unnaturalistic dualism, that might not be a compelling argument against keeping consciousness in spacetime, but it surely gives us an incentive to explore the prospects for an atemporal account of the stream of consciousness.

5.4 An atemporal solution

The heart of Dainton's theory, and the key to his solution of the paradox of the stream of consciousness, is the idea that two experiences can have an experiential constituent in common. This is what allows a stream of consciousness to comprise phenomenally overlapping experiences without involving the sort of phenomenal excess that generates the problem of repeated contents.

Now, in Dainton's theory, the part that two phenomenally overlapping experiences have in common is a *temporal* part—a stage or phase of a diachronically complex experience. But there is no obvious reason why the part must be temporal. In particular, nothing in the introspectible, phenomenal quality of our experience requires us to think of the part as temporal.

If there is a phenomenally integrated collection of experiences, each member of which shares an *atemporal* part with the member (or members) of the collection with which it phenomenally overlaps, then we also have phenomenal integration without repeated contents. Provided that we can make sense of the idea of a sharing of experiential parts between two experiences that transcend time and space, there is nothing to prevent us from formulating a version of Dainton's theory of the stream of consciousness that is consistent with an atemporal view of experience.

Things have properties; at least, this is true of most things, and all experiences. It often happens that two things have a property in common: a cherry and a fire hydrant both have the property of being red, a cow and a cactus both have the property of being alive, etc. As these examples demonstrate, two things can share a property without sharing a part. Although the cow and the cactus both instantiate the property of vitality, the cow's instantiation of the property is a different instantiation of it from the cactus'. Between the two of them, the cactus and the cow instantiate vitality twice: the property gets instantiated by the cow, and also, and independently, by the cactus.

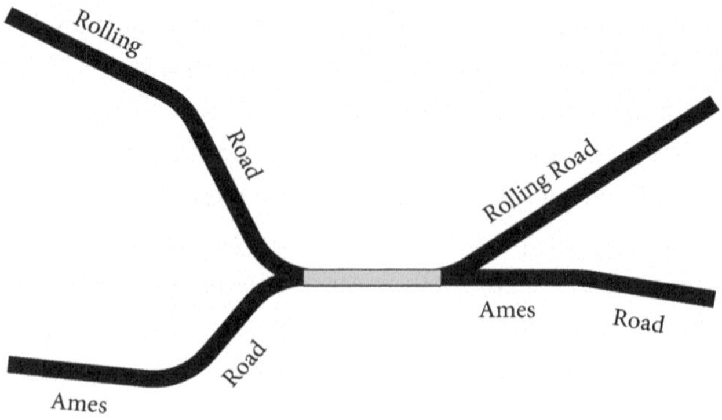

Figure 5.5 Concurrent roads.

But what if two things share not only a property, but a property *instantiation*? That is, what if one thing's instantiation of a given property is one and the same instantiation of it as another thing's? Consider two roads that are paved except for over a stretch where they run concurrently (see Fig. 5.5). Here, Rolling Road and Ames Road both instantiate the property of being partially unpaved. But unlike the case of the cow and the cactus, the roads' instantiation of this property is a *single* instantiation: both roads instantiate the property, even though (or even if) the property gets instantiated only once.

This example suggests a general criterion for part-sharing: two things, x and y, share a part if and only if the set of x's instantiations of properties intersects the set of y's instantiations of properties; that is, if and only if there is a property that both x and y instantiate, even though (or even if) the property gets instantiated only once. As a special case of this, two phenomenally-individuated conscious experiences share a part just in case the qualia-instantiations that characterize one of the experiences include a qualia-instantiation that also characterizes the other experience.[14]

[14] In the case where x and y have only a finite number of properties, x and y have a part in common if and only if the number of property-instantiations involved in x's existence plus the number of property-instantiations involved in y's existence is greater than the number of property-instantiations involved in the existence of both x and y.

The necessary and sufficient conditions for part-sharing stated in the text do not imply an analysis of parthood. This is because part-sharing typically involves a common instantiation of a property of the form, *is partially* ϕ. For example, Utah and the United States share the Great Salt Lake as a common part, but neither Utah nor the United States instantiates the property of lakehood or (arguably) saltiness. They do, however, both instantiate the property of being *partially* salty, and would both instantiate it even if that property were instantiated only once. By contrast, Utah and Israel do not have any part in common, despite the fact that they both instantiate the property of being partially salty; if the property of being partially salty were instantiated just once, then either Utah would not contain the Great Salt Lake, or Israel would not contain the Dead Sea.

What are the instantiators of qualia—the bearers of phenomenal properties? This is a question I'll have something to say about in later chapters. For now, let us see how far we can account for the stream of consciousness without making any assumptions about the specific nature of the instantiators of qualia (or even about the necessity of instantiators for instantiations).

Consider a set *S* of qualia with the following properties: (1) the qualia in *S* are coinstantiable, and (2) if a quale Q is in *S*, then, for each type of quale such that a quale of that type must be instantiated in order for Q to be instantiated, *S* contains exactly one quale of that type.

Such a set is what we can call a *phenomenal character*.

A conscious experience is a coinstantiation of all (and only) the qualia in a phenomenal character.[15]

For two properties to be coinstantiated is for there to be something that has both properties. As I use the term "coinstantiate," coinstantiation is a highly generic relation among properties. When I say that a property *F* is coinstantiated with a property *G*, I do not assume (a) that the coinstantiator of *F* and *G* is something over and above an instantiation of properties (as on a substrate theory of particulars), or (b) that the coinstantiator of *F* and *G* is *not* something over and above an instantiation of properties (as on a bundle theory of particulars), or (c) that the coinstantiator of *F* and *G* instantiates *F* and *G* simultaneously (a typical human life instantiates both infancy and adulthood), or (d) that the coinstantiator of *F* and *G* instantiates *F* and *G* successively (an object can simultaneously possess mass and charge), or (e) that the coinstantiator of *F* and *G* instantiates *F* and *G* in time (the number 5 instantiates both oddness and primeness).

I am looking at a ripe tomato; my experience has a certain phenomenal hue, call it Phenomenal Red 5 (PR-5, for short). But a set that has PR-5 as its only member is not a phenomenal character. An experience as of something red must also be an experience as of something with certain luminance properties (after all, I can see the same tomato under different lighting levels). So in order for an experience to have phenomenal redness (or PR-5 in particular), it must also have some phenomenal luminance. Since a set of qualia that contains only PR-5 does not contain any luminance quale, it does not contain a quale of each type that must be instantiated in order for PR-5 to be instantiated.

Typically, if not always, a given quale features in more than one possible experience. PR-5 must be coinstantiated with *some* luminance quale, but there is no luminance quale that PR-5's instantiation requires. PR-5 might be coinstantiated

[15] In Leibnizian terms, a phenomenal character is a phenomenological "complete notion."

with Phenomenal Luminance 22, or PL-3, or PL-57. (Analogously, a square must have *some* area, even though there is no area that a square must have.)

Suppose now that we have a set of qualia that includes PR-5, PL-22, and, for each other type of quale such that PR-5's instantiation requires an instantiation of a quale of that type, a quale of that type. Call this set V. Suppose, somewhat controversially (see below), that this set includes only visual qualia (a hue quale, a luminance quale, a visual size quale, etc.).

A coinstantiation of the qualia in V is an example of what I shall call an *atomic experience*. An atomic experience is a coinstantiation of the qualia in a phenomenal character that has no phenomenal character as a subset.

Suppose we add an auditory quale to V—say, Phenomenal Rumbling (of the kind that characterizes hearing thunder). The resulting set of qualia is not the phenomenal character of any possible experience because it fails to satisfy condition (2) for phenomenal characterhood. Phenomenal Rumbling cannot be instantiated without the instantiation of some loudness quale (for example); since the V + Phenomenal Rumble set does not contain any loudness quale, it does not satisfy (2) and is not a phenomenal character.

To get a phenomenal character again, we can add more auditory qualia to the set (a volume quale, perhaps qualia associated with auditory distance and orientation, etc.) until the set again satisfies (2). Call the resulting set AV.

AV is a phenomenal character, and a coinstantiation of the qualia it contains is an experience—but not an atomic experience. The experience is non-atomic, because its phenomenal character (AV) has two subsets, each of which independently satisfies conditions (1) and (2) for phenomenal characterhood: an auditory subset and a visual subset.

An experience having AV for its phenomenal character is what I shall call a *molecular experience*. A molecular experience is simply a coinstantiation of the qualia in a phenomenal character that can be partitioned into two or more phenomenal characters.

Two experiences intersect if there is a quale-instantiation that both contain. Atomic experiences can intersect: one atomic experience's instantiation of PR-5 might be one and the same instantiation of PR-5 as another atomic experience's instantiation of PR-5 (and not just an instantiation of the same quale; viz. PR-5).

Assuming there are molecular experiences, they, too, can intersect. Unlike intersecting atomic experiences, the intersection of two molecular experiences can (but need not) be an experience in its own right. For example, a molecular experience with phenomenal character AV might intersect a molecular experience having phenomenal character A′V, with the same visual qualia as AV, but different auditory qualia. In that case, the two molecular experiences have an experiential part in common: an experience with phenomenal character V.

What are atomic experiences like? This is the same as asking what our simplest experiences are like—a question to which there may be no simple answer (hence the controversy alluded to earlier).

One possibility is that an atomic experience is a multi-modal phenomenal field, characterized by different types of qualia (visual, auditory, sensory-somatic, cognitive, etc.). This suggestion goes along with a kind of experience-holism, according to which each phenomenal aspect of an experience somehow reflects every other phenomenal aspect of the experience. On this view, the visual character of a deaf person's experience of waves crashing onto a shore differs from the visual character of a non-deaf person's experience of the same thing. An audiovisual experience of the waves does not consist of two phenomenal "tracks," either of which we might turn off without affecting the other. Rather, if one were to "turn off the sound" (remove the auditory qualia), one would also thereby alter the *visual* phenomenology of the experience, effectively removing the experience's original visual qualia too.

Another possibility is that an atomic experience is typically a mono-modal experience. This goes along with the idea that someone watching and listening to the sea could have had exactly the same visual phenomenology, even if he hadn't had any auditory experience at all (and the same auditory phenomenology, even if he hadn't had any visual experience).

If atomic experiences are mono-modal, it must be possible for them to combine to form multi-modal experiences; any difficulty in constructing our multi-modal experiences out of mono-modal elements would weigh in favor of a holistic conception of atomic phenomenology, according to which *all* experiences are (phenomenologically rich) atomic experiences.

However, the case against mono-modal atomic experience is not as strong as some make out. If I see someone near me strike a bell with a hammer, I expect to hear a sound. If I don't hear a sound, I'll be surprised. Thus, it is true that the experience I have if I see and hear the bell differs from the experience I have if I see but do not hear it, and in a non-auditory (or at least not purely auditory) way. But the difference is just the sense of surprise at not hearing a sound. If we imagine watching a video recording of someone repeatedly striking a bell, it may be that the audio track of the recording cuts in and out: sometimes footage of the bell being struck is accompanied by the sound of the bell, sometimes it plays silently. If we imagine the conscious experiences that we have in this situation, where we have no expectation of hearing (or not hearing) a sound, I think it becomes more plausible to say that the visual quality of our successive experiences (as of a bell being struck, or a recording thereof) is the same, whether or not the visual phenomenology is accompanied by auditory phenomenology.[16]

Fortunately, we do not have to take a stand on the phenomenal character of atomic experiences (mono- versus multi-modal) in order to give an atemporal account of the stream of consciousness. We can define a stream of consciousness as a series of experiences $e_0, e_1, e_2, e_3, \ldots e_n$, such that for every positive integer $i \leq n$, experience e_i phenomenally overlaps and logically intersects e_{i-1}. In other words, a stream of consciousness is a phenomenally and constitutionally integrated group of conscious experiences, where phenomenal integration is understood in the usual way (in terms of phenomenal overlap), and constitutional integration is understood as a logical intersection of groups of qualia-instances, rather than as a partial coincidence of spatiotemporally extended phenomenal processes.[17]

This atemporal account of the stream of consciousness, like Dainton's, solves the paradox of the stream of consciousness. That is not surprising, since the atemporal theory is closely modelled on Dainton's theory, with one notable difference. Dainton's theory requires diachronically complex experiences, which, as we have seen, implies the violation of an otherwise perfect symmetry of nature. Since the atemporal account of the stream of consciousness does not presuppose the existence of diachronically complex experiences, it avoids this implication. If this isn't an argument for removing conscious experience from time, it is at least an argument for entertaining the possibility that conscious experience does not occur in time (or spacetime)—an argument for taking the atemporal position seriously.

5.5 Phenomenal sorites

I have argued that an atemporal theory of the stream of consciousness is, from a phenomenological standpoint, at least as plausible as the best available temporal theory. A key premise of this argument is that phenomenal overlap is a requirement for phenomenological streaming. The existence of this requirement emerged through a comparison of cases in which a stream of consciousness continues, to

16 Something similar happens in the McGurk effect, where the quality of one's auditory experience of another person's speech depends on the quality of one's visual experience of that person's mouth. Here, too, the bearing of one modality on another, while strong, would appear to be less than logically necessary, and thus less than what it would have to be to support a holistic conception of atomic phenomenology. (For the McGurk effect, see McGurk and Macdonald (1976).)

17 Alternatively, we can define a stream of consciousness in terms of the ancestral of the overlapping-and-intersecting relation. This is the relation—call it R^*—in which an experience x stands to an experience y if, and only if, x belongs to every set Z such that (1) y is a member of Z, and (2) everything that overlaps and intersects a member of Z is a member of Z. We can then define a stream of consciousness as any set of experiences S such that (1) the members of S all stand to one another in R^*, and (2) for all x, if there is a member of S to which x bears R^*, then x belongs to S.

similar cases in which one stream ends and another begins (the kitchen and the forest, etc.). Before concluding this chapter, we must consider a challenge to this requirement, and to the idea that phenomenal overlap plays any essential role in the stream of consciousness.

The challenge arises from the fact that the phenomenal overlap requirement entails that a stream of consciousness that ends with a different phenomenal color from that with which it begins must contain at least one experience as of color-change, whereas it seems possible—or so it is claimed—for a stream that includes no appearance of color-change to begin with an appearance of one color, and end with an appearance of another color.

We can state this challenge as an argument:

C1. If a stream of consciousness is necessarily a series of phenomenally overlapping experiences, then a stream of consciousness whose first and last experiences have different phenomenal colors must contain at least one experience as of color-change.

C2. A stream of consciousness whose first and last experiences have different phenomenal colors need not contain any experience as of color-change.

C3. So, a stream of consciousness is not necessarily a series of phenomenally overlapping experiences.

It is hard to see how C1 could fail to be true. If none of the experiences making up a given stream of consciousness is an appearance of color-change, then the only way for the stream to contain experiences instantiating different phenomenal colors would be to contain an experience as of one unchanging color followed by an experience as of a distinct unchanging color. But it is hard to see how there could be phenomenal overlap between two such experiences: one would be an experience as of this-color-and-then-still-this-color, while the other would be an experience as of that-color-and-then-still-that-color.

So it seems we should accept C1. But what about C2?

This premise of the argument ostensibly receives support from experiments in which one watches a slowly changing scene without noticing that it is changing. The interesting thing about these cases is that the phenomenal quality of the experience that you have at the end of the experiment differs from that of the experience that you had at the beginning.

For example, you might be watching a screen on which a solid expanse of color is projected. The color changes gradually over the course of a minute or two, say from pink to purple. The experimenters do not inform you of this. Typically, subjects in this situation do not notice any change in the phenomenal color of their experience over the course of the experiment. Yet, if shown the first and last images projected on the screen in direct succession, they uniformly report a difference in the phenomenal colors of the attendant experiences. This makes it very plausible to suppose that the first and last visual experiences they had during the experiment also differed in phenomenal color.

There is one interpretation of these experiments that we should avoid. On this interpretation, one has, in a situation such as that described above, a stream of consciousness consisting of experiences as of change that we cannot introspect as such: instantiations of the property of phenomenal color-change that are immune to introspection. This is not a good interpretation, since, as argued in Chapter 1, this kind of phenomenology *is* introspectible, at least in principle, by whoever has it.

A better objection to C2 is that the streams we have in the described experiments *do* contain introspectible appearances of change. It's just that we tend not to introspect them when we first try the experiment. Once we are shown what changed, we can introspect experiences as of color-change when trying the experiment again. This makes it plausible to suppose that the phenomenal color-change was there in the original stream of consciousness too, just that we needed to have our attention directed to it.

At least, this is my experience. One can, of course, imagine an extended version of the experiment, where the same color change (of the image on the screen) takes place much more slowly, perhaps over the course of an hour or more. In such a case, it might well be that I am incapable of introspecting any appearance of color-change, yet end in a different phenomenal color-state from that in which I began. If so, the explanation is that in such a case, I have more than one stream of consciousness while viewing the screen. That is, at some point (or multiple points) during the experiment, I have an experience that does not phenomenally overlap the experience that occurs before it. So, for example, at some point, I have an experience as of $pink_1$-then-more-$pink_1$ followed by an experience as of $pink_2$-then-more-$pink_2$, without any experience as of $pink_1$-then-$pink_2$:

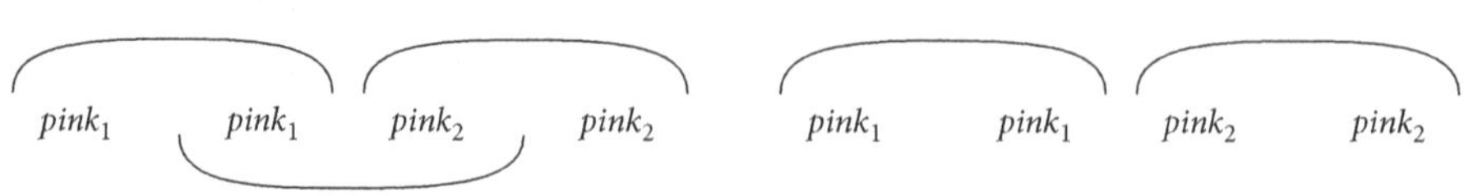

(a) phenomenal overlap (b) no phenomenal overlap

Figure 5.6 Single stream versus separate streams.

With reference to this figure, the idea is that in the extended experiment, the subject's sequence of experiences contains one or more subsequences like 5.6(b), where there is no experience as of $pink_1$-then-$pink_2$, but rather a first stream of consciousness that ends with an experience as of pink$_1$, and a second that begins with an experience as of pink$_2$.

Now, for this to be plausible, we have to assume that the subject occasionally misremembers the phenomenal color of his recent experience as being the same as the phenomenal color of his current experience, or at least fails to remember his recent experience as having had a different phenomenal color from his present experience. Otherwise, it would occur to him that the color was changing, whereas this does not occur to him at all in the course of the experiment.

So the proposal is a kind of error theory. But it is an innocuous one. The cognitive failure it posits is of a familiar kind: misremembering something as having happened one way when really it happened a slightly different way. The picture is this: over the course of the very-slow-change experiment, you have a sequence of experiences: E_1, E_2, E_3, . . . up to E_n, for some rather large n. We may suppose that most (maybe almost all) pairs of adjacent experiences in the sequence are complex experiences in their own right; for example, we may suppose that E_1 and E_2 are parts of a single, complex experience; that E_2 and E_3 are parts of a single, complex experience; and that for most j, E_{j-1} and E_j are parts of a single, complex experience. But at least one pair of adjacent experiences—say, E_{606} and E_{607}—is such that there is no experience to which they both belong. E_{606} is the end of one stream of consciousness, and E_{607} the beginning of another.

You could not possibly notice via introspection that E_{606} and E_{607} did not belong to the same experience, or to the same stream of consciousness, since the most that one can apprehend in an act of introspection are experiences that belong to a single, complex experience as parts or phases. As a matter of logical possibility, you could recall, while having E_{607}, that the phenomenal color of your preceding experience (E_{606}) was slightly different from the phenomenal color of E_{607}. But it would be surprising if you did recall this, given how small the difference in phenomenal properties would be, and how imperfect memory is. If, while you are attentively staring at a cup, the cup moves to the right by some small fraction of an inch that is just barely within your powers to perceive, the motion is something you will probably notice. But if you momentarily glance away from the cup and then back to it (or blink), and if the cup moves the same fraction of an inch to the right while you are not looking at it, you will almost certainly *not* notice that it has moved. Likewise, you will almost certainly not realize that there is a difference between the phenomenal colors of E_{606} and E_{607} if these experiences are not phases of a single experience.

Anyway, if there were no lapse of memory, then there would have to be an unnoticeable appearance of change; but surely, if I could fail to notice that one of my occurrent experiences was an appearance of change, I could all the more easily fail to remember that one of my experiences was unlike another in some subtle respect. Solving the phenomenal sorites by appeal to a failure of memory is more modest, and more plausible, than solving it by appeal to a failure of cautious introspection.

We normally associate breaks in the stream of consciousness with episodes of losing or regaining consciousness, and it is true that a paradigm case of streamal termination is falling asleep, and a paradigm case of streamal initiation is waking up. But these are not the only ways in which a stream can begin or end. God might create or annihilate a conscious human being mid-stream, or my present mental state might have been immediately preceded by the mental state I actually was in an hour ago. Breaks between the streams of consciousness one has while staring for hours at the colored screen can be subtle versions of this kind of streamal discontinuity.[18]

5.6 Conclusion

This brings us to the end of the chapter, and the end of the first half of the book. Let us summarize our main findings so far.

Consciousness has several features that encourage us to apply temporal predicates to it. First, there is phenomenal change. This is an intrinsic, introspectible feature of individual experiences, which temporalist and atemporalist theories of consciousness can account for equally well, at least as far as consistency with the evidence of introspection is concerned.

Second, there is phenomenal simultaneity. Again, we have found that both temporalist and atemporalist accounts of this are possible, although not without significant repercussions for temporalism (in the form of a likely commitment to Russellian dualism).

Finally, there is the flow or stream of consciousness revealed by memory. Dainton accounts for the stream of consciousness in terms of a phenomenal and constitutional integration of experiences. This also makes sense on an atemporal conception of experience, where constitutional integration takes the form of a sharing of non-spatial, non-temporal elements (rather than partial temporal

[18] Realistically, staring at a uniformly-colored screen for hours at a stretch would likely result in non-veridical experiences as of color-change, just as listening to an unvarying auditory tone for a long period of time tends to result in non-veridical experiences as of tonal variation. A number of such non-veridical experiences might take one from an initial appearance of one color (e.g. pink) to a final appearance of another color (e.g. purple). But even if one has no non-veridical appearances of change while staring at the screen for an hour, the phenomenal colors of one's first and last experiences in the experiment can differ, by virtue of belonging to different streams of consciousness, as in Figure 5.6(b).

coincidence). Indeed, we have found that an atemporal theory of the stream of consciousness has at least one advantage over the temporalist theory, which is its compatibility with the symmetry of fundamental natural laws.

Phenomenological change, presence, and flow: these are the loosely-speaking temporal features of conscious experience. They are the sort of features that Poincaré took to conflict with a scientific understanding of time and space, and that Russell believed to complicate the relation between perception and physics.

Whether Poincaré was right to think that there is no room in physical time for conscious experience is debatable. I have argued that we must say that Poincaré *was* right, if we respect the unity of consciousness and the symmetry of Nature. What we have established without qualification is that there is no compelling phenomenological reason to think that conscious experience *does* occur in spacetime. Phenomenological change, presence, and flow are features that conscious experience can have, even on an atemporal conception of experience, and on rather more satisfactory terms than on a temporalist conception, at least as far as the phenomenology of the situation is concerned.

Almost everyone thinks that conscious experience changes. Almost everyone thinks that an ordinary visual impression of a banana is simultaneously yellowish and curvy. And almost everyone thinks of the stream of consciousness as a process unfolding in time.

I submit that these inclinations, or the misconceptions into which they unconsciously harden, have blinded generations of philosophers to the possibility that conscious experience is not a temporal phenomenon at all—and consequently, to the possibility that consciousness might serve as a suitable basis for the metaphysical reduction of temporal phenomena, including time (or spacetime) itself.

Whether such a reduction is feasible remains to be seen, but the mere possibility of it is so intriguing that one would have expected its potential to have been explored in depth by now. The reason why this has not happened is that practically no one has ever thought that there was a coherent possibility here to explore. Now that we know there is, we can explore it ourselves. That is the plan for the rest of this book.

6

Idealism

There is much here that is correct and close to my own view. But it is expressed paradoxically. For it is not necessary to say that matter is nothing, but it is sufficient to say that it is a phenomenon, like the rainbow; and that it is not a substance, but the resultant of substances, and that space is no more real than time, that is, that space is nothing but the order of independently existing things, just as time is the order of things standing in relations of dependence. True substances are monads, that is, perceivers. But the author should have gone farther, to the infinity of monads, constituting everything, and to their pre-established harmony.

Gottfried Wilhelm Leibniz, *Handwritten notes on Berkeley's Treatise*

6.1 Introduction

If consciousness does not exist in spacetime, how does it relate to things that do? The most promising answer to this question is that spacetime and its contents relate to conscious experience by supervening on facts about it. For the only obvious alternative to this answer is to say that the relationship between consciousness and the contents of spacetime is causal, and it is very hard to see how there could be a causal relationship between two domains, only one of which existed in time (or spacetime).[1]

[1] To say that the facts about spacetime and its contents—"physical facts," for short—supervene on the phenomenological facts is to say that the obtainment of the actually obtaining phenomenological facts is logically sufficient for the obtainment of the actually obtaining physical facts. As will become apparent later, I am actually going to argue for more than this. I am going to argue (in Chapter 8) that we can give an *analysis* of each individual physical fact into phenomenological terms. For now, there is no need to belabor the distinction between supervenience and analyzability (or reducibility).

The proper way to understand supervenience claims is actually a matter of some debate, but the debate revolves around relatively minor issues of formulation, discussion of which would only distract from the central aims of the present study. Nothing about my central position hangs on which of the more precise definitions of supervenience we go with, or whether we formulate idealism in terms of "grounding" instead of supervenience. For relevant discussion, see Chalmers (1996a), 38–41, Jackson (1998), 1–27, Schaffer (2009), and Stoljar (2010), 127–43.

It might be worth mentioning that one of the main motives for refining the simple notion of supervenience is that most *physicalists* who want to frame their theory in terms of supervenience

The idea that the physical facts supervene on the phenomenological facts is not new. It is the central idea of all idealism. In this chapter, we survey the options available to an idealist, and classify idealist theories according to how they approach the reduction of the physical (or spatiotemporal) to the experiential.

The simplest idealist theory is one that equates a physical object or event with a combination of experiences—an apple, for instance, with a combination of sensations of color, texture, motion, odor, etc. This is how Berkeley seems to think of it:

> By sight I have the ideas of light and color with their several degrees and variations. By touch I perceive, for example, hard and soft, heat and cold, motion and resistance, and of all these more and less either as to quantity or degree. Smelling furnishes me with odors; the palate with tastes, and hearing conveys sounds to the mind in all their variety of tone and composition. And as several of these are observed to accompany each other, they come to be marked by one name, and so to be reputed as one thing. Thus, for example, a certain color, taste, smell, figure and consistence having been observed to go together, are accounted one distinct thing, signified by the name *apple*. Other collections of ideas constitute a stone, a tree, a book, and the like sensible things; which, as they are pleasing or disagreeable, excite the passions of love, hatred, joy, grief, and so forth.[2]

Berkeley's proposal has several major shortcomings.[3]

First, by equating physical things with combinations of conscious experiences (or "ideas"), Berkeley commits himself to the view that we eat conscious experiences. After all, we eat apples, and apples are, according to Berkeley, nothing

also want to recognize the logical possibility of "ghosts" or "epiphenomenal ectoplasm"—conscious experiences that are wholly independent of any physical state of affairs. If ghosts are logically possible, then presumably there is a logically possible world that is just like the actual world in all respects *except* for that it contains some additional "ghostly" experience. But if such a world is logically possible, then it is false that the obtainment of the actually obtaining physical facts is logically sufficient for the obtainment of the actually obtaining phenomenological facts: in a ghost world, there obtain all the physical facts that actually obtain, but not the actually obtaining phenomenological fact (as we may suppose it to be) that there are no ghostly experiences. But it is not clear that idealists who want to frame their theory in terms of supervenience face any analog of the ghost problem. For, as we shall see, the very considerations that give us a reason to favor idealism also give us a reason to deny the logical possibility of any physical analog of ghosts: physical entities that are wholly independent of any phenomenological state of affairs. In any case, if supervenience idealists do have a problem analogous to the ghost problem, they can either avail themselves of one of the solutions to this problem that physicalists have developed, or shift from talk of supervenience to talk of grounding.

2 Berkeley (1710/1982), 23.

3 Howard Robinson categorizes Berkeley as a "theistic phenomenalist": see Robinson (1994), 226–8. Below, I define phenomenalist theories as idealist theories that do not posit any non-experiential origins for experiences. By this definition, Berkeley is not a phenomenalist, since he holds that our experiences (or at least very many of them) originate from God.

but combinations of conscious experiences. But if we know anything about conscious experiences, it is that we do not eat them.[4]

Second, Berkeley's idealism rules out the possibility of unperceived things, let alone imperceptible ones, since it says that all things are essentially constituted by perceptual experiences.

Third, since Berkeley equates an apple with a certain combination of sensations, it follows that if that combination of sensations were the only thing in the universe, then the apple would exist. This, despite the fact that if the universe had contained additional sensations, these additional sensations might have revealed the apple-sensations to have been deceptive (part of a dream, hallucination, or whatever).

Any defensible idealism must differ from Berkeley's in two key respects.

First, it must recognize the foundational importance of phenomenological *potential*. A defensible idealism must not try to reduce physical facts to facts about actual experience alone, but to facts about some kind of potential for experience.

Second, it must take a holistic approach to the logical construction of physical reality. An idealist must construe the existence of an apple as a fact about the sum total of *all* experiences, rather than a fact involving only the sort of experiences one has when one takes oneself to perceive an apple.

All of history's more sophisticated idealist metaphysics exhibit both of these features. We discuss holism below. But first, let's classify idealist theories according to how they conceive of phenomenological potential. This is not the only useful classification, but it is, I think, the most illuminating (see Fig. 6.1).

I define idealism as the view that facts about spacetime and its contents supervene on broadly phenomenological facts, but not vice versa. By "broadly phenomenological facts," I mean facts about conscious experience or potential for conscious experience. While different idealists appeal to different kinds of potential, all think of a potential for an experience as requiring more than the mere logical possibility of the experience, but less than the experience's actual existence.

No idealist thinks of phenomenological potential in terms of *physical* causes, or potential causes, of experience. An idealist denies that experiences have physical causes, or that anything physical explains anything about conscious experience. That is one of the things that distinguishes idealism from all forms of physicalism, and most forms of dualism.[5]

[4] For Berkeley on eating ideas, see Berkeley (1710/1982), 38. Of course, there are a lot of other equally improbable things we also do with ideas, on Berkeley's view.

[5] Idealism is incompatible with physicalism twice over: once, because physicalists deny that the physical supervenes on the phenomenological, and once again because physicalists affirm that the phenomenological supervenes on the physical. Dualism is also incompatible with idealism, but only because dualists (like physicalists) deny that the physical supervenes on the phenomenological.

The most conservative idealism is Kant's. According to Kant, experiences have partly non-mental origins, arising from the interaction of minds and non-mental, non-physical "things-in-themselves" (or "noumena," as Kant sometimes calls them). To say that an experience "arises from" interactions between some noumena and a mind is to say that something about how that mind relates to those noumena explains why the experience occurs. (Likewise, to say that an experience arises from a mind is to say that some fact about a mind explains why the experience occurs.) The explanation need not be causal, though it typically is understood in broadly causal terms.

According to Leibniz and Berkeley, experiences arise from minds alone: there are no noumena in Leibnizian or Berkeleyan metaphysics. The main difference

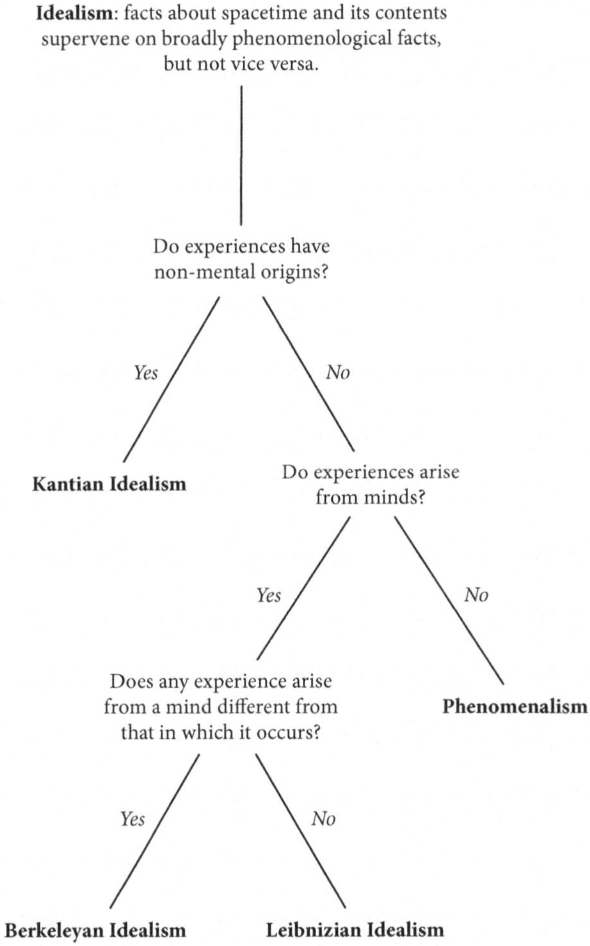

Figure 6.1 Varieties of idealism.

between Berkeley and Leibniz is that according to Berkeley, conscious experiences (or at least many of them) arise from interactions between finite minds like ours and a divine Mind, while according to Leibniz, each experience originates solely from the operations of the mind in which it occurs.

Mill stands apart from these other idealists, in that he denies that anything explains the occurrence of any conscious experience, unless possibly it is the occurrence of another conscious experience. This is the phenomenalist position. According to a phenomenalist, the existence of one experience (or collection of experiences) can give a high probability to the existence of another experience, but this is just a brute fact that has no explanation in terms of the experiences' relationship to some mind, noumenon, or physical state of affairs. If the existence of a given conscious experience has some explanation, it is only in the sense that some other experiences make the existence of that experience probable.

I have not given a name to the class of theories falling under the node marked by the question: "Do experiences arise from minds?" We could call this class of theories "narrow idealism" to emphasize the fact that only these theories aspire to reduce physical reality to something purely mental. Kant is not a narrow idealist, since he understands the physical world in terms of potential interactions between conscious minds and non-mental (but also non-physical) things-in-themselves.

I have also not given a name to the position corresponding to the node immediately above Berkeley and Leibniz, marked by the question, "Does any experience arise from a mind different from that in which it occurs?" We could call theories falling under this node "mental substrate idealism."

We explore these various idealist theories below. But first we should say something about the holism that must characterize any defensible idealism, and does characterize the idealist metaphysics of Leibniz, Kant, and, to a less certain extent, Mill. This is the holism that favors reducing the existence of an individual physical object or event to some fact about the totality of all conscious experiences (or potential conscious experiences), rather than just the sort of experiences that might characterize a perception of that object or event. In Leibniz, this holism takes the form of a harmony or agreement among monads' mental lives:

> It is true that the things that happen in the soul must agree with those that happen outside the soul; but for this it is sufficient that those things that happen in one soul correspond both among themselves and with those things that happen in any other soul; and there is no need to posit something outside of all souls or monads.[6]

[6] Leibniz (1712/2007), 257. In the same letter, Leibniz writes: "I regard the explanation of all phenomena solely through the perceptions of monads agreeing among themselves . . . to be useful for a fundamental investigation of things. In this way of explaining things, space becomes the order of coexisting phenomena, as time is the order of successive phenomena, and there is no absolute or

> Just as the same city viewed from different directions appears entirely different and, as it were, multiplied perspectively, in just the same way it happens that, because of the infinite multitude of simple substances [i.e. monads], there are, as it were, just as many different universes, which are, nevertheless, only perspectives on a single universe, corresponding to the different points of view of each monad.[7]

Kant is never as explicit as Leibniz about the need to take a holistic approach to the idealization of the physical world, but it is hard to see how Kant could avoid a holistic idealism, given his holism about the physical universe that is to be idealized.[8]

Mill does not develop his metaphysical position as fully as Leibniz or Kant, and it is unclear that he saw the importance of taking a holistic approach to the analysis of physical things. But, as we shall see, he proposes to equate the existence of an individual physical object with the truth of a conjunction of counterfactual conditionals of the form, "If such-and-such experiences were to occur, such-and-such other experiences would also occur," and there is at least a suggestion of holism in Mill's recognition that for such an equivalence to hold, the conjunction must contain a large, possibly even infinite, number of conjuncts.[9]

Although phenomenological holism takes different forms in different idealist theories, the basic idea is the same in all cases. Let's pin this idea down with an analogy.

Suppose I hand you an envelope full of loose photographs. Sifting through these, you find that you can fit them together—not literally, like a jigsaw puzzle, but thematically—as pictures of a birthday party. One of the photographs in particular you take to be a picture of a partygoer.

Now I hand you a thicker envelope with even more photographs. Combining these with the contents of the first envelope, you find you can fit them all together as depicting a party scene in a live stage production. Considered in the context of this wider class of photographs (the first plus the second envelope), you no longer interpret the aforementioned photograph as a picture of a guest, but an actor.

I might go on to hand you a suitcase full of photographs, which, in combination with the pictures from the envelopes, you find you can piece together as documenting an elaborate birthday party held on the stage of a theater leased expressly

spatial nearness or distance between monads. To say that they are crowded together in a point or disseminated in space is to employ certain fictions of our mind when we willingly seek to imagine things that can only be understood."

7 Leibniz (1714/1989a), 220.

8 See Kant (1781/1998), 378–9, and the discussion in Langton (1998), 56–63.

9 See Mill (1865/1979), 179; more on Mill's "sensation-conditionals" below.

for that purpose. Now you interpret the photograph mentioned earlier as a picture of a party guest again.

Just as a picture is best interpreted as depicting a guest or an actor, depending on the totality of pictures that we consider along with it, an experience is best interpreted as a perception of one thing or another (or nothing), depending on the totality of experiences that we consider along with it.

What makes one interpretation of a totality of experiences better than another? Well, different interpretations may be better for different purposes, but when I speak of "the best interpretation" of a totality of experiences, I mean: best for the purpose of making the world of things in time and space intelligible to us. There are various respects in which one interpretation of a totality of experiences might be better than another, relative to this purpose. Without pretending to give a complete catalogue of them, the main desiderata are, I think, the following.

First, there is the desideratum of Unity. Other things being equal, an interpretation of a totality of experiences, according to which the totality includes perceptions only of things existing in a single, common spacetime, is better than an interpretation of the same totality as including some perceptions of things existing in one spacetime, and other perceptions of things existing in other spacetimes. Other things being equal, a one-world interpretation is better than a many-worlds interpretation. This is connected with the fact that many-worlds interpretations are extremely easy to come by. For example, you could lazily posit as many spacetimes as there are experiences, and construe each experience as a perception of something in its own special spacetime.[10]

The requirement of Unity is not very stringent. Just as it is easy to construe *n* experiences as perceptions of things in *n* distinct spacetimes, it is easy to construe *n* experiences as perceptions of things in *n* locations in a single spacetime. For example, we could interpret my experiences as perceptions of one planet, your experiences as perceptions of a distinct but similar planet, and so forth for each person's experiences, while supposing all these planets to occur in a single spacetime.

In addition to Unity, we should therefore introduce a desideratum of Economy. Other things being equal, one interpretation of a totality of experiences is better than another, if it is more economical than the other; that is, if it interprets the totality as including perceptions of fewer things (objects or events) than the other interpretation. If I have an experience as of a landscape, and you have a very

[10] It could turn out that the best interpretation of a totality of experiences is a many-worlds interpretation. In that case, and other things being equal, an interpretation that posits fewer worlds is better than one that posits more.

similar experience as of a landscape, it is better, other things being equal, to interpret our experiences as perceptions of a *single* landscape (perhaps observed from different vantage points), rather than as perceptions of distinct but similar landscapes.

Of course, things are not always equal. It may be that we really are viewing distinct but similar—even visually indistinguishable—landscapes. Only a holistic consideration of the totality of all potential experiences can decide whether this is the best way to interpret our experiences. For example, it depends on whether we can interpret the totality of potential experiences as including perceptions of traveling along a non-looping path from one landscape to a landscape visually indistinguishable from it.[11]

A third main desideratum of an interpretation of a totality of experiences has to do with how well it organizes the things that it interprets the totality as including perceptions of (spatiotemporally located objects and events) into an intelligible system. This is the desideratum of Order. For example, it is possible to interpret our experiences as including perceptions of bodies that obey different laws of motion, depending on their situation relative to the Earth. Following Aristotle, we could interpret our experiences as including perceptions of sublunary objects tending to move toward the center of the Earth at a constant velocity, and superlunary objects tending to orbit the Earth in concentric circles. Alternatively, we can interpret our experiences in such a way that we construe all perceptions of bodies as perceptions of objects obeying Newton's inverse square law of gravitation. The latter interpretation does a better job of satisfying the desideratum of Order than the former.

While Unity, Economy, and Order seem to be the main desiderata for an interpretation of a totality of experiences (relative to the goal of interpreting it in a way that makes the physical world intelligible to us), there are probably other desiderata as well. I have not attempted to give a complete account of these. Nor have I attempted to weigh the relative importance of the desiderata that

[11] Actually, this case suggests that in all strictness, it might be best to state idealist theories in terms of streams of consciousness, rather than just individual experiences. Suppose that I am looking down an empty stretch of highway in Texas. The experiences I have might be indistinguishable from the ones that you have while looking down an empty stretch of highway in Nevada. I have suggested that we can distinguish this scenario from one in which we are looking down the same stretch of highway, by reference to the existence of a possible experience as of traveling from my location (in Texas) to yours (in Nevada). The problem with this suggestion is that there is no such thing as a conscious experience as of traveling from Texas to Nevada (at least, not for us). A solution is to switch from consideration of experiences to consideration of streams of consciousness. For while there can be no such thing, for us, as an experience as of traveling from Texas to Nevada, a stream of conscious experience connecting appearances of a road in Texas to appearances of a road in Nevada is perfectly possible.

I have discussed. The important thing here is that we really do recognize some such desiderata: we judge, or are disposed to judge, some interpretations of a given collection of experiences as doing a better job than others of making the physical world intelligible to us. We distinguish between perceiving a snowball, and imagining, hallucinating, or dreaming about a snowball. We interpret some of our experiences as revealing something about our environment, others as failing to do so. If my attempt to outline the criteria that guide these interpretations has been crude or inept, I invite others to do better. It is enough for present purposes that the criteria are implicit in our actual practice, ready to be characterized.

Now let's take a closer look at historical idealist theories that appeal to some kind of phenomenological potential as the logical foundation of physical reality.

6.2 Kant's metaphysics of the thing-in-itself

According to Kant, the existence of an apple reduces to the existence of things-in-themselves with suitable experience-causing powers. The apple is not to be equated with the noumenon that has the power to cause experiences as of an apple, nor is it to be equated with the experiences themselves (à la Berkeley). All we can say is that, necessarily, the apple exists, with whatever physical properties it has, if and only if noumenal reality harbors the potential to cause suitable conscious experiences—experiences interpretable, in the context of all noumenalistically possible experiences, as perceptions of an apple.[12]

Kant often says that we can know nothing positive about the nature of things-in-themselves. However, in context, it seems clear that he means that we can know nothing positive about the nature of things-in-themselves *besides that they have certain powers to cause experiences.* After all,

> In whatever way and through whatever means a cognition may relate to objects, that through which it relates immediately to them, and at which all thought as a means is directed as an end, is intuition. This, however, takes place only insofar as the object is given to us; but this in turn is possible only if it affects the mind in a certain way. The capacity (receptivity) to acquire representations through the way in which we are affected by objects is called sensibility . . . [A]ll thought, whether directly or indirectly, must ultimately be related to intuitions, thus, in our case, to sensibility, since there is no other way in which objects can be given to us.[13]

[12] Thus, it is a mistake, according to Kant, to think that our appearances provide a basis for *identifying*, for example, a water droplet with any "transcendental object" (i.e. noumenon): Kant (1781/1998), 186–7.

[13] Kant (1781/1998), 155; see also Kant (1781/1998), 375.

Kant's views about the relationship between minds and noumena are actually rather unclear. In some places, he insists that the concept of causation applies only within the domain of experience—the implication being that it does *not* apply to the noumena.[14] However, I see no way to make sense of Kant's metaphysics without supposing that the noumena can give minds experiences, in some sense of "give," and I see no way to understand this giving except in broadly causal terms. Perhaps Kant would draw a distinction between two kinds of causation: a sui generis kind that holds between a noumenon and a mind it stimulates, and an empirical kind that holds between one idealistically-constructed object and another (e.g. a tennis ball and a tennis racquet). Settling these interpretive questions is beyond the scope of the present study; for now, I assume that the relation between a noumenon and a mind to which it gives an experience is a causal or quasi-causal relation of asymmetric dependence (of mind, or mental state, on noumenon, but not vice versa).[15]

It is also fairly clear that Kant thinks that things-in-themselves are *not* mental entities (e.g. minds, or a divine Mind). For one thing, Kant thinks that equating noumena with minds results in an overly mentalistic conception of physical reality. This comes out most clearly in Kant's criticism of Leibniz. According to Kant, Leibniz's decision to think of monads as the sole origins of their respective experiences collapses the distinction between physical and mental reality—between a "phenomenon" and a corresponding "appearance" or "representation" of that phenomenon.[16]

Furthermore, we have seen that Kant takes conscious experiences—and so, presumably, the minds in which they occur—to exist in time, whereas he insists that things-in-themselves are neither spatial nor temporal.[17]

The defining feature of Kantian idealism is its appeal to non-mental causes, or potential causes, of experiences. Like all idealist theories, it is best to understand Kantian idealism holistically: the apple is what exists if and only if realization of noumenal potential to cause experience tends to yield sum totals of experience that are interpretable as including perceptions of the apple. For Kant, the existence of a physical thing comes to the fact that things-in-themselves are disposed to

[14] See Kant (1781/1998), 275 and the surrounding discussion.

[15] Arguably, Kant's "transcendental apperception" is supposed to be just such a sui generis relation between minds and noumena; see Kant (1781/1998), 249–51, and also Kant (1781/1998), 434–5.

[16] I do not think that this is a fair criticism of Leibniz; if it were, we could equally well accuse Kant of collapsing the distinction between physical and noumenal reality. Relevant sources are Kant (1781/1998), 368 and Kant (1781/1998), 373–4; see Langton (1998), 70–89 for discussion. Kant's criticisms of Berkeley are harder to construe, but he certainly objected to Berkeley's appeal to a divine Mind as the source of (veridical) human experience. See Kant (1955), 58.

[17] See Kant (1781/1998), 163–5 and Kant (1770/1894), 59 (the latter quoted in Section 3.1).

interact with minds so as to give rise to experiences that present a picture of a physical world containing that thing.

6.3 Leibniz's monadology

In the Kantian picture, there are minds (receptive to noumena) and noumena (potent to minds). The existence of a physical thing reduces to the fact that the noumena have certain powers: powers to give the minds experiences that collectively lend themselves to interpretation as including perceptions of a world containing that thing. The actual existence of minds is unnecessary for the existence of physical things; all we need for that is noumena that have the *potential* to produce suitable experiences in minds, and they could have this potential even if there were not, as a matter of fact, any minds around to be affected.

Let us now modify this picture, replacing noumena with minds, and powers to cause experiences with tendencies to *have* experiences. Instead of equating an apple with a tendency for noumena to give minds experiences interpretable as perceptions of a world containing an apple, let us equate it with a tendency for minds to have experiences interpretable in that way.

These modifications leave us with a view that is basically identical to that of Leibniz.[18]

And finally you ask, "why are these appearances produced in me or in other true substances [i.e. monads]?" I say: those that follow are produced from the preceding appearances in accordance with metaphysical and mathematical laws of eternal truth. But the reason why there are any such appearances at all is the same as the reason why the universe exists. For you can easily see that the simple substances can be nothing but the sources and

[18] As I interpret him, without any claim to scholarly authority. The only really controversial feature of my interpretation, I think, is its minimization of the role of God. Leibniz sometimes describes God as the creator (and programmer) of the monads. Taken at face value, these comments make Leibniz into a kind of proto-Berkeleyan, with God functioning as the ultimate source of all experience. But it is also possible to interpret Leibniz's God as a personification of the harmony of the monads, where this harmony is taken as the fundamental thing. That is how I interpret Leibniz.

In any event, God gets into Leibniz's system more on a technicality than as a vital working part. For Leibniz, God is just whatever explains the existence and harmony of the monads; God's existence is guaranteed, Leibniz thinks, by the principle of sufficient reason, according to which every contingent state of affairs has some explanation. As many have argued, and as I will argue in Chapter 7, this principle is false. But that does not result in the collapse of Leibniz's philosophy. It just allows us to remove a decorative base that gave the edifice no real support. Instead of saying that the buck stops with God, we say that it stops with the harmony of the monads. By contrast, Berkeley's system makes no sense at all without God, or at least becomes highly indeterminate between a Leibnizian monadology and a Millian scheme of permanent possibilities of sensation (of which more below).

Discussions of "pre-established harmony" are scattered throughout Leibniz's later writings: see the many references to pre-established harmony in Leibniz (1989), Leibniz (2007), and Leibniz (1998).

principles of so many series of perception unraveling themselves in order, expressing the same universe of phenomena with the greatest order and variety.[19]

For Leibniz, a mind—or "monad," as he calls it—is a sort of phenomenological Turing machine, characterized by a disposition to have certain experiences upon having certain other experiences. Leibniz characterizes the monad as an "automaton" (what we would today call a computer) that runs a program that takes the monad's current phenomenal state as input, and yields its next phenomenal state as output:

In fact, one can say that the soul [i.e. monad] is a very exact immaterial automaton . . . [T]he soul, even though simple, always has feelings composed of several simultaneous perceptions; which for our purposes has the same effect as if it were composed of parts, like a machine. For, in conformity with a law of order which exists in perceptions as much as in motions, each preceding perception influences succeeding ones.[20]

Different monads have different phenomenological dispositions—they are governed by different programs for converting experiential input into experiential output. Depending on its program and its initial input, a given monad will generate streams of consciousness belonging to a single mental life like yours or mine. If we consider all the streams of consciousness that monads generate (or would generate, given initial inputs that led to the generation of coherent streams of consciousness), we find that we can interpret the streams as containing perceptions of various physical states of affairs. These states of affairs constitute physical reality. An apple's existence is equivalent to the fact that the monads tend to generate experiences that are collectively interpretable as including perceptions of a physical world that contains, among other things, an apple.[21]

Ontologically, Leibniz's idealism is more economical than Kant's, since it posits only minds (monads), rather than both minds and things-in-themselves. In Leibniz's view, there is no need to think of conscious experiences as having extra-mental causes at all. The monadology is idealism even on a narrow construal of "phenomenological facts" as not including facts about putative extra-mental causes or potential causes of experiences, but only facts about the intrinsic properties of minds, experiences, and streams of consciousness.

[19] Leibniz (1704/1989), 184. [20] Leibniz (1698/1998), Sections 10–12: 206.

[21] If monads really are Turing machines, then a monad must always give the same phenomenal state as output for the same phenomenal state as input. In that case, phenomenally indistinguishable experiences could only occur in me within indistinguishable streams of consciousness; for example, I could not have an overwhelming pain that abated and (on some other occasion) a phenomenally indistinguishable pain that intensified. This would probably not have bothered Leibniz though: since he affirms the principle of the identity of indiscernibles, he would probably say that there couldn't be such a thing as two phenomenally indistinguishable states of mind.

So Leibniz's idealism is more ambitious than Kant's. But an even more ambitious idealism is possible.[22]

6.4 Mill's permanent possibilities of sensation

Leibniz equates physical reality with a propensity for monads to have certain experiences—experiences that collectively yield to interpretation as perceptions of a common world of things in time and space—and he conceives of this propensity in terms of phenomenological inputs and outputs. This conception is consistent with our usual ways of thinking about propensities (potentials, dispositions, tendencies, etc.). We say that power tends to corrupt, meaning that for an input of power, we usually get an output of corruption. Fire has the power to burn marshmallows; input: fresh marshmallow, output: burnt marshmallow. Edward has artistic potential; input: the right training and encouragement, output: a good artist.

Now, when we speak of there being some power, potential, or tendency, there is often a presumption of something that *has* the power, potential, or tendency. Usually, we think of a power or potential as a *mediated* input-output architecture. This is how Leibniz thinks of monads.

But although we usually think of powers and potentials as involving entities that have those powers or potentials, we do not always think of them this way. Consider the tendency for a body at rest to remain at rest. We *can* think of this in terms of a body's having a certain input–output architecture; that is, we can think of a body as a Turing machine that returns an output of *continue to stand still* for an input of *standing still now*. But we don't have to think of it this way. In fact, in this case we do *not* refer the tendency to any further feature of the body that we say has it. It is not because a body has some inner structure that it implements the "stand still" program. This is clear from the fact that even the simplest (and presumably structureless) parts of a body also implement the program. Rather, we say that bodies just do tend to behave this way; there is no reason to posit a mediator between input and output.

So, even though Leibniz himself depicts minds as mediators of phenomenal inputs and phenomenal outputs, there is no obvious reason why a modern-day idealist should retain this aspect of Leibniz's theory. Instead, we might think of phenomenological potential as a kind of *unmediated input–output regime*—an input–output regime that manifests itself in the form of fundamental phenomeno-

[22] Leibniz's theory is an example of what Howard Robinson calls "analytical phenomenalism"; see Robinson (1994), 226–8. In my taxonomy, Leibniz is an idealist, but not a phenomenalist.

logical regularities, rather than in the form of suitably programmed immaterial automata.

This is John Stuart Mill's position. According to Mill, the existence of a given physical thing reduces to the truth of one or more ungrounded counterfactual conditionals concerning conscious experiences. Following Skorupski, let's call these conditionals "sensation-conditionals."[23]

A sensation-conditional is a proposition of the form: *If sensations of such-and-such a kind were to occur, then sensations of such-and-such another kind would occur*. An example is the proposition that if there were to occur an appearance of something sweet and banana-shaped, there would occur an appearance of something sweet, banana-shaped, and yellow. Another example is the proposition that if there were to occur an appearance of something white and crescent-shaped against a black starry background, there would occur an appearance of something white, crescent-shaped, and luminous.[24]

These two sensation-conditionals are, by Mill's account, first steps towards analyses of the propositions that ripe bananas are yellow, and that the Moon is luminous, respectively. A full analysis of these propositions (or any other proposition to the effect that some physical state of affairs obtains) would have to be much more complex, consisting of a conjunction—perhaps a very long conjunction—of sensation-conditionals. A conjunction of true sensation-conditionals that does intuitively entail the obtainment of some physical state of affairs (the existence of some physical object, or the occurrence of some physical event) is what Mill calls a Permanent Possibility of Sensation. The physical world is, according to Mill, nothing but a (vast, possibly infinite) collection of Permanent Possibilities of Sensation.[25]

[23] See Skorupski (1994), 114. Valberg calls the conditionals "Phenomenalist conditionals"; see Valberg (1992), 161.

[24] Skorupski says that a sensation-conditional has the form: "If such and such sensations were to occur, then such and such other sensations would occur with a given degree of probability." I have replaced Skorupski's "such and such sensations . . . such and such other sensations" with "sensations of such and such a kind," so as not to prejudge questions about the proper individuation of sensations, or the question of how many experiences one has at any given time. I have also omitted Skorpuski's "with a given degree of probability," since this qualification does not seem to me to have any basis in Mill's writings. On the contrary, Mill holds that sensation-conditionals describe "conditional certainties." Mill (1865/1979), 180. This is not to say that the idea of bringing probability into the picture isn't a good one—I believe it is—it's just that I see no clear precedent for such a move in Mill. One advantage of using (as Skorupski does) sensation-conditionals with probabilistic consequents is that such counterfactual conditionals are just about the only ones that can be true, according to Alan Hájek; see Hájek (2014).

[25] Mill's theory is an example of what Howard Robinson calls "sceptical phenomenalism," although there is nothing particularly skeptical about it; see Robinson (1994), 226–8.

Mill's phenomenalism is even simpler than Leibniz's idealism, since it does not even include minds as fundamental posits. Minds exist, sure enough, on Mill's view, but physical states of affairs reduce to facts about *experiences,* independently of whether or how those experiences constitute entire minds. When it comes to minds, Mill is basically a bundle-theorist: the monad, as an irreducible phenomenological substance, has no place in his metaphysics.[26]

6.5 Conclusion

Idealism, in any form that can hope to succeed, consists of two claims. First, some kind of potential for conscious experience is *necessary* for the existence of anything physical. Second, the same kind of potential for conscious experience is *sufficient* for the existence of anything physical.

Both of these claims face important objections.

The main objection to the necessity claim is that no potential for experience is necessary for the existence of *imperceptible* physical phenomena, including (arguably) all of the most fundamental physical phenomena (elementary particles, fields, forces, etc.).

The main objection to the sufficiency claim is that we can conceive of a situation in which there exist few, if any, ordinary physical things, but in which there does exist all the phenomenological potential that actually exists, only harbored in an Evil Demon, or a supercomputer connected to some envatted brains.

These challenges confront all idealist theories; we consider them in detail in Chapter 9. Specific versions of idealism (Kantian, Leibnizian, and Millian) invite further objections specific to each version, though not to idealism in general. This is especially true of phenomenalism, the alleged shortcomings of which we examine in the next chapter.[27]

[26] Other prominent, albeit in some cases wavering, phenomenalists in the Millian tradition include Bertrand Russell, Frank Ramsey, and A.J. Ayer; see Russell (1910–11) (interpreting "sense-data" in light of the exchange at the very end of Russell (1918)), Ramsey (1925), 291, and Ayer (1946–7), respectively.

[27] The next three chapters may be read as an extended response to David Armstrong's comprehensive criticism of phenomenalism in Armstrong (1961), 47–79.

7

Phenomenalism: A first appraisal

> The conception I form of the world existing at any moment, comprises, along with the sensations I am feeling, a countless variety of possibilities of sensation: namely, the whole of those which past observation tells me that I could, under any supposable circumstances, experience at this moment, together with an indefinite and illimitable multitude of others which though I do not know that I could, yet it is possible that I might, experience in circumstances not known to me. These various possibilities are the important thing to me in the world.
>
> J.S. Mill, *An Examination of Sir William Hamilton's Philosophy*

7.1 Introduction

Of all the varieties of idealism, phenomenalism is the most ambitious. This makes phenomenalism much more intriguing than other idealist metaphysics, but it also means that phenomenalism faces objections that other forms of idealism do not have to contend with. Two of these objections stand out as particularly pressing.

The first objection is that phenomenalism makes an unacceptable mystery of the fact that conscious experience exhibits the order and coherence that it does in fact exhibit. What gives this objection its force is that the phenomenalist, unlike other idealists, cannot call on the causal or quasi-causal powers of minds or things-in-themselves to explain why conscious experience tends to take the orderly form it does.

The second objection is that phenomenalism unacceptably implies that we never perceive anything but our own conscious experiences, since it implies that the physical universe is a construction out of conscious experiences, and the only experiences to which one has any kind of direct cognitive access are one's own.

We address these objections in Sections 7.3 and 7.4, respectively. But first, it is worth establishing that there is at least one respect in which phenomenalism is superior to other narrow idealisms, like Berkeley's and Leibniz's. This is that, contrary to popular belief, phenomenalism avoids commitment to a mind-dependent picture of physical reality.

7.2 Mind-dependence

One of the most common criticisms that a phenomenalist encounters is that his position makes physical reality too dependent upon minds. Thus, Armstrong:

> The argument may be put briefly. The nature of the merely empirically *possible* is determined by what is actual. For the Phenomenalist, in a 'no-mind' universe nothing is actual, therefore nothing [and so, no experience] is empirically possible, therefore nothing exists at all.[1]

Armstrong then adds that "[s]trictly, this argument does not refute the Phenomenalist. He can simply accept its conclusion. But it is a conclusion that the modern Phenomenalist is not likely to relish."[2]

What, exactly, is the problem for phenomenalism supposed to be here? Armstrong assumes, correctly, that phenomenalism proposes to reduce physical facts to potentials for experience, understood as something like Mill's permanent possibilities of sensation. But in order for such potentials to exist, Armstrong reasons, they must have some categorical basis—there must be something by virtue of which any given potential for experience exists. In particular, the existence of a potential for experience presupposes the existence of some *mind* that is capable of having the experience (even if it never actually does have it). Thus, if there were no minds, there would be no potential for experience, and therefore, according to phenomenalism, no physical reality. Or so the argument goes.

Armstrong's argument relies on the claim that in order for there to be an empirical possibility for some state of affairs (as opposed to a mere logical possibility for it), there must exist some actual thing or things by virtue of whose nature that empirical possibility exists. But if Mill's permanent possibilities of sensation are examples of empirical possibilities, then it is hard to see why we should accept this claim. (If permanent possibilities of sensation are *not* examples of empirical possibilities, then Armstrong's argument is irrelevant to Mill's position.) Even if the universe did not actually contain any physical phenomena, it might have been true that physical changes would have conserved energy, had there been physical changes. Likewise, even if the universe did not actually contain any minds or experiences, it might have been true that experiences would have presented a picture of an objective physical universe, had there been experiences.

For Mill, a physical world devoid of conscious experience would simply be a world in which various propositions of the form, *Had there been such-and-such experiences, then there would have been such-and-such other experiences*

[1] Armstrong (1961), 56–8. See also Smart (1963), 22–3.
[2] Armstrong (1961), 58.

were true, but in which all propositions of the form, *There are such-and-such experiences* were false. Since for Mill minds themselves are nothing more than certain potential patterns of experiences—potential streams of consciousness, or suitable collections thereof—there is nothing to prevent Mill from admitting the possibility of a physical world devoid of both minds and conscious experience.

So on Mill's view, the existence of a physical world does not depend at all on the existence of anything mental. It depends on the *possibility* of conscious experiences, and on various facts about what sorts of experiences would occur were certain other sorts of experiences to occur, but this possibility and these facts do not depend for their reality on the actual existence of any mind or conscious experience.

This is in contrast to Leibniz's view, which is vulnerable to Armstrong's criticism (not that this would have perturbed Leibniz). Unlike Mill, Leibniz does ground potentials for experience in the phenomenological input–output architecture of the monads; so, if there were no minds (and hence no monads), there would be no potential for experience, and therefore, by Leibniz's lights, no physical world.

There is a connection between Armstrong's criticism of phenomenalism and the accusation that phenomenalism is a kind of "anti-realism" about physical reality. This accusation is either false or harmless, depending on what those who make it mean by "anti-realism."

If anti-realism about physical things is the view that physical things are not real—that there aren't really any physical things—then the accusation is false. A phenomenalist does not deny the reality of physical things: he simply proposes a specific theory about the *nature* of physical things. A phenomenalist who equates physical objects with potentials for experience does not deny the reality of physical objects, any more than an economist who equates risks with potentials for loss denies the reality of risks.

If anti-realism about physical things is understood as the view that physical things depend for their existence on the existence of minds, then some, but not all, idealists are anti-realists about physical things. As we have seen, Leibniz does think that facts about physical reality reduce to facts about actually existing minds (or monads); so, by this definition of "anti-realism," Leibniz is an anti-realist. But Mill is not an anti-realist by this definition, since minds play no part in his logical construction of the physical world. Anyway, it is misleading to call even Leibniz's theory "anti-realist," since it does not deny the existence of anything that has any pre-theoretical claim to reality.

If, finally, we define "anti-realism" about physical things as the view that the existence of physical things reduces to facts about conscious experience broadly construed (including facts about potentials for conscious experience), then phenomenalism is, of course, by its very nature an anti-realist position. But this is an entirely defensible anti-realism—or so I am in the process of arguing. Really, it is just misleading to use "anti-realism" with this third meaning; it would be better simply to speak of reductionism in this context.[3]

7.3 Ultimate mysteries

We now return to the two objections that specifically target phenomenalism, beginning with the charge that phenomenalism makes a mystery of the fact that we have the specific conscious experiences that we do, and that these experiences occur in specific patterns. This is in supposed contrast to a view according to which our experiences have underlying grounds or causes, where it is these that account for the observed patterns in experience.[4] Thus, according to J.J.C. Smart,

> the rather incoherent stream of our sense impressions is readily understood if they are thought of as due to our interaction with an objectively existing physical world, but on the phenomenalist theory it is a huge accident that they hang together in the way they do.[5]

Although Smart describes the stream of consciousness as "rather incoherent," it is actually the relative *coherence* of our streams of consciousness that seems to call for explanation—that is, their coherence, relative to what one would expect if our experiences occurred at random.

[3] "Realism" is one of those words that philosophy would probably be better off without. For a detailed discussion of its multifarious uses and misuses, see Dummett (1963).

[4] A Kantian can explain the patterns we find in our experience by reference to the causal structure of a noumenal level of reality, and a Leibnizian can explain the coherence of our experiences by reference to the input–output architecture of monads. As for why minds actually have experiences, and why they possess the specific input–output architecture they do rather than some other architecture, the Leibnizian cannot say. He can account for *some* of a monad's experiences by reference to the monad's program and some of the monad's other experiences. But assuming that a monad does not have an infinite number of experiences, we cannot account for *all* of a monad's experiences by reference to its program and prior experiences, and even if a monad does have an infinite number of experiences, the Leibnizian cannot explain why the monad has some experiences rather than none. (Leibniz is quite explicit about all of this; see the first quote in Section 6.3.) Not that this puts Leibniz at a disadvantage to Kant: Kant can't explain why minds are affected by the specific noumena (things-in-themselves) that affect them, or why the noumena have the particular experience-causing powers they do. (Concerning the former point, the Kantian may be tempted to say that a particular mind interacts with a particular noumenon by moving into the vicinity of that noumenon. However, motion and proximity are spatial concepts which, like all spatial concepts, the Kantian seeks to understand in terms of noumenal potentials. It would be circular for the Kantian to explain noumeno-mental interactions in terms of spatiotemporal proximity.)

[5] Smart (1963), 25.

Smart's objection to the phenomenalist position has a ring of plausibility, but it does not, I think, bear up to careful scrutiny. It is true that the phenomenalist offers no explanation for why our experiences occur the way they do. Nor could he, consistently with his theory, according to which the patterns that characterize conscious experience have the status of fundamental regularities—fundamental, and therefore inexplicable. But why should this be considered a problem for phenomenalism?

One reason for thinking it is a problem is that there is no logical necessity for our experiences to exhibit the coherence that they do, which suggests that there must be some explanation for the fact that they do exhibit such coherence—some reason why experience exists in these particular patterns, rather than in different patterns, or not at all.

This is an appeal to the principle of sufficient reason, according to which every contingent state of affairs has some explanation (although not necessarily an explanation that *we* can grasp). While this principle has a certain intuitive appeal, it is, I think, demonstrably false.[6]

Suppose it is a contingent fact that there obtain all, and only, those states of affairs that do, actually, obtain. That is, suppose that it is a contingent fact that *p* and *q* and *r* and . . . , where for every way that things are, but might not have been, the statement that they are that way is equivalent to the statement that *p*, or equivalent to the statement that *q*, or equivalent to the statement that *r*, or . . .

According to the principle of sufficient reason, this fact has some explanation. The explanation cannot invoke any contingent state of affairs, since that would be circular: it would be an attempt to explain the existence of a totality (the totality of all contingent states of affairs) by reference to one of its own parts.

Thus, if something explains why there obtain the specific contingent states of affairs that do, this must be a non-contingent, necessary state of affairs.

Moreover, it must be a necessary truth that this necessary state of affairs (call it *N*) explains the totality of contingent states of affairs. For suppose that it is only contingently true that *N* explains that totality. Then the fact that *N* explains the totality is itself part of the totality. In that case, *N* would explain not only the totality, but also the fact that *N* explained the totality. But that cannot be. If my car's battery is dead, what explains this may be that I left the headlights on. But the fact that I left the headlights on does not explain the fact that the battery is dead due to my having left on the headlights. If someone asks, "Why does the fact that you left your lights on explain the fact that your battery is dead?" the correct answer is not "Because I left my lights on." (Maybe the correct answer is: because the lights

[6] The argument that follows essentially reproduces that of Inwagen (2009), 150–3.

use a lot of battery power, which does not get replenished when the engine isn't running.)

So, if the total contingent state of the world has some explanation, the explanation is that its existence is necessarily explained by a necessarily obtaining state of affairs.

But whatever a necessarily obtaining state of affairs necessarily explains is itself necessary. (If it is necessarily the case that *p*, and if the fact that *p* necessarily explains the fact that *q*, then it is necessary that *q*—for if the fact that *q* failed to obtain, then it would either not be the case that *p*, or not be the case that *p* explained *q*.)

So, if there is an explanation for the fact that there obtain just those contingent states of affairs that do actually obtain, then the fact that those contingent states of affairs obtain is a necessarily obtaining fact. This, however, is a contradiction in terms. It cannot be that the obtainment of these contingently obtaining states of affairs is necessary.

What led to the contradiction?

We derived it from two claims: (1) that it is only contingently true that the world is the way it is in contingent respects (i.e. only contingently true that *p* and *q* and *r* and . . .); and (2) that the principle of sufficient reason is true. Since (1) is undeniable, we must reject the principle of sufficient reason.

There is one historically influential objection to the foregoing reasoning that we should address. This is an objection to the claim that if something explains the fact that things are exactly the way they contingently are, it cannot be among the ways things contingently are (since that would make the explanation circular). Against this, Hume famously argues that it *is* (or at least might be) possible to explain the total contingent state of the world other than by reference to something that is not part of it, by explaining each *part* of contingent reality by reference to another part of contingent reality. After all, as Hume points out,

> In such a chain, too, or succession of objects, each part is caused by that which preceded it, and causes that which succeeds it. Where then is the difficulty? But the WHOLE, you say, wants a cause. I answer, that the uniting of these parts into a whole, like the uniting of several distinct countries into one kingdom, or several distinct members into one body, is performed merely by an arbitrary act of the mind, and has no influence on the nature of things. Did I show you the particular causes of each individual in a collection of twenty particles of matter, I should think it very unreasonable, should you afterwards ask me, what was the cause of the whole twenty. This is sufficiently explained in explaining the cause of the parts.[7]

[7] Hume (1779/2007), 65–6.

Hume's suggestion is that we might explain each contingently occurring event by reference to some other contingent event that causes it. This presupposes that history contains an infinite number of events, but, as far as we know, it does. Assuming that explaining each part of a whole is sufficient for explaining the whole, it follows that it is, or might be, possible to explain why the world is contingently the way it is, without referring to anything beyond the realm of contingency.

But the assumption is false. The example of the twenty particles establishes only that you have explained a whole if you have explained each of its parts *in such a way that you do not explain each part by reference to another part.* The point does not generalize to cases that fail to satisfy the italicized condition, as the following thought-experiment should make clear.[8]

Suppose we discover a spherical object drifting in interstellar space. We ask: why does this object exist? It turns out that the object consists of a series of nested spherical shells. For each shell, there is another, thinner shell nested inside of it. Furthermore, each shell is exactly twice as thick as the next shell in (see Fig. 7.1, where *r* is the radius of the sphere).

Someone now offers the following explanation of the sphere's existence: "The outermost shell, with a thickness of $r/2$, was generated by the next shell in (with thickness $r/4$). The latter shell was in turn generated by the next shell in (with thickness $r/8$). And so on, ad infinitum. It took 1 second for the $r/4$ shell to generate the outermost shell, half a second for the $r/8$ shell to generate the $r/4$ shell, a quarter of a second for the $r/16$ shell to generate the $r/8$ shell, etc., for a total of two seconds for the sphere to materialize."

Has the existence of the sphere now been explained?

No. We still don't know why this whole process of concentric generation occurred. This, despite the fact that we have explained why each *part* of the sphere exists (by reference to another part of the sphere).

Likewise, even if we could, as Hume suggests, explain each contingent event by reference to some other contingent event, and in this way provide an explanation for every *part* of contingent reality, this would not yield an explanation of the existence of contingent reality as a whole.

The argument against the principle of sufficient reason therefore stands. Not all contingent states of affairs admit of explanation. There are *ultimate mysteries.*

Since there are ultimate mysteries, every metaphysical system must leave something ultimately mysterious. But different metaphysical systems may imply different ultimate mysteries, and some may ask us to tolerate more mystery than

[8] I am not the first to criticize Hume for overgeneralizing on this score: see Rowe (1975), 264.

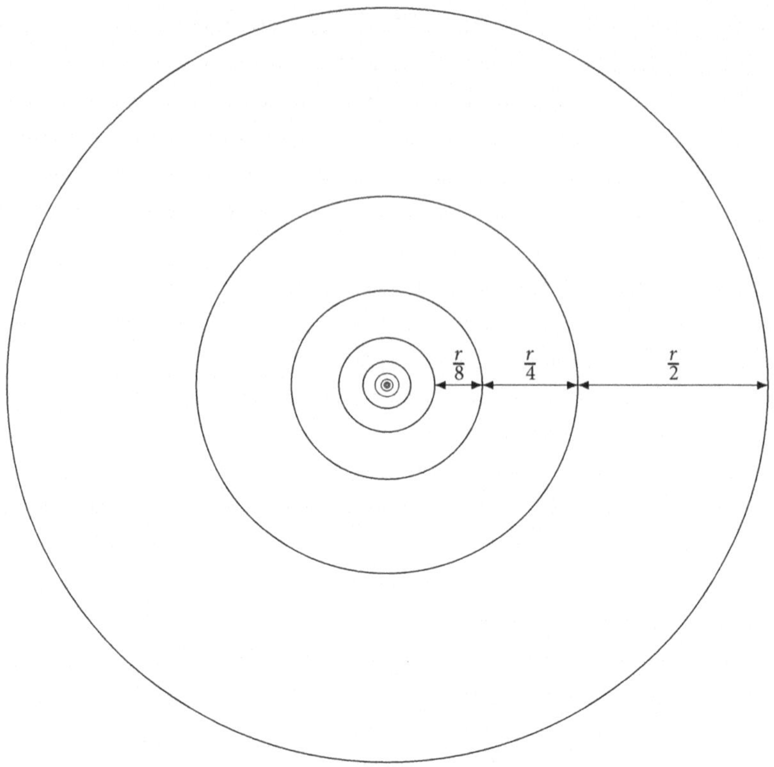

Figure 7.1 The interstellar object, in cross section.

others. The real question is whether the things that a given metaphysics leaves ultimately mysterious are things that we are prepared to accept as such.[9]

[9] It is not obvious that there is any fact that *all* metaphysical systems deem ultimately mysterious. The best candidate for such a fact is probably the fact that there exists anything at all.

But according to some metaphysicians, every possible world is real. These "modal realists" consider it no mystery that our world exists—in this respect, our world does not differ from any other possible world. Nor, according to the modal realist, is there anything mysterious about the fact that there are such things as possible worlds: the existence of possible worlds is a necessary fact that does not call for explanation. Like the necessitarian, the modal realist dissolves the mystery of the existence of our world by saying that all worlds exist; it is just that the modal realist says that there are an infinite number of worlds, whereas the necessitarian says that there is just one. (For modal realism, see Lewis (1986a), 128–33.)

Still, even the modal realist must recognize the existence of ultimate mysteries. From the modal realist point of view, it is mysterious that just one world exactly like ours exists—or, if there is more than one world exactly like ours, why there exist precisely as many such worlds as there do. It is true that we have no need to suppose that there is more than one world just like ours, but that just means that if there is only one world just like ours, it is a mystery why there are no more such worlds than we need suppose.

Phenomenalism implies that there is no explanation for the fact that our experiences exhibit the patterns and regularities that they do. But this does not mean that our experiences occur at random. A phenomenalist does not believe that it is an improbable fluke or statistical miracle that our experiences admit of interpretation as perceptions of an orderly physical universe. But nor need he believe that something *explains* why our experiences occur the way they do. For it is open to him to hold that it is a fundamental law of nature—or of experience—that they occur that way.

The point is missed often enough to bear emphasis. "Caused versus random" is a false dichotomy. Nothing (as far as we know) *causes* bodies to conform to the inverse square law of gravitation. ("The force of gravity makes them conform" is just a confusing way of saying that they do conform.) Likewise, nothing *causes* physical phenomena to behave only in ways that conserve energy. That physical things behave only in energy-conserving ways is not a consequence of some further fact about those things: it is a fundamental regularity in the things' behavior.

Yet, we do not conclude from this that the behavior of physical things is random; quite the opposite. Nor do we consider it a miracle that physical phenomena exhibit intelligible patterns and lawlike regularities—or if we do consider it a miracle, it is a miracle we have learned to accept.[10]

According to physicists, it is a fundamental law of the physical world that in it energy is always conserved. Well, according to a phenomenalist, it is a fundamental law of the *phenomenal* world that its constituent experiences are interpretable as perceptions of a physical world in which energy is always conserved. The main difference between the laws, according to the phenomenalist, is that the latter law is more fundamental. The relative coherence of conscious experience—coherence

[10] Contrast these remarks with Chalmers' take on the hypothesis that "I do not receive inputs from anywhere in the world. Instead, I have random, uncaused experiences. Through a huge coincidence, they are exactly the sort of regular, structured experiences with which I am familiar." Chalmers (2010a), 479. According to Chalmers, this "Chaos Hypothesis" is a skeptical hypothesis, because

> on the Chaos Hypothesis, there is no causal explanation of our experiences at all, and there is no explanation for the regularities in our experience. In all of the previous cases [involving envatted brains], there is some explanation for these regularities, though perhaps not the explanation that we expect. One might suggest that as long as a hypothesis involves some reasonable explanation for the regularities in our experience, then it will not be a global skeptical hypothesis. (Ibid.)

But we might accept this last suggestion, without accepting the converse, that a hypothesis is non-skeptical only if it involves some reasonable explanation for the regularities in our experience. For, as we have seen, it is doubtful that "what is crucial is that on the Chaos Hypothesis, there is no causal explanation of our experiences." Rather, what is crucial is that on the Chaos Hypothesis, experiences occur at random, with no tendency or propensity to exhibit the regularities that they do. This is not the same as saying that nothing *explains* why the experiences exhibit those regularities. They might exhibit the regularities by occurring in a way that is neither random nor explicable.

relative to what one would expect of randomly occurring experiences—is a reflection of the fact that experience is subject to certain fundamental laws, or exhibits certain fundamental regularities, rather than of experiences being caused in certain ways.[11]

A Kantian might agree that there are such things as fundamental laws and regularities, but claim that Mill is mistaken to locate these laws or regularities at the level of experience. But what does the Kantian offer as an alternative? Fundamental laws and regularities at the level of things-in-themselves (or the level of noumenomental interactions). And how is this an improvement on Mill's position? It might be an improvement if some further details about the precise nature of these noumenal regularities were forthcoming, but the status of the noumena in the Kantian theory guarantees that no such details will ever emerge. A Berkleyan or Leibnizian might suggest that it is more satisfying to *account* for the regularities in our experience than to leave those regularities unaccounted for. But again, the chain of explanations stops somewhere, and it is unclear what advantage we gain by positing an independent object or class of objects (God, monads, or whatever) as the final link in the chain, rather than stopping at the experiences themselves.

There is no escaping the existence of fundamental, inexplicable regularities. The question is where we locate them in our metaphysics. For the Kantian, the fundamental regularities are certain relationships between minds and rather mysterious things-in-themselves. For the Berkeleyan or Leibnizian, the fundamental regularities take the form of certain volitional or computational dispositions of minds. For the phenomenalist, the fundamental regularities are simply regularities in conscious experience itself.

7.4 The veil of ideas

The second objection to phenomenalism mentioned earlier was that it implies that the only things we can perceive are our own conscious experiences. Concerning this objection, we may be brief.

[11] Leibniz makes the point about laws of experience taking priority over laws of physics when he says: "What I take to be the indivisible or complete monad is the substance endowed with primitive power, active and passive, like the 'I' or something similar, and not those derivative forces which are continually found first in one way and then another . . . The forces which arise from mass and velocity are derivative and belong to aggregates [of monads], that is, to phenomena. And when I speak of the primitive force that persists, I don't understand the conservation of total motive power about which we were once concerned, but the entelechy that always expresses that total force, as well as other things. And indeed, derivative forces are only the modifications and resultants of the primitive forces." Leibniz (1703/1989), 176.

Far from implying that we perceive only our own conscious experiences, phenomenalism does not even imply that the only things we *directly* perceive are our own conscious experiences.

David Armstrong asks: "What is the *direct* or *immediate* object of awareness when we perceive?", and applies the label "direct realist" to theories according to which the answer is: typically an ordinary physical object or event. Phenomenalism, according to Armstrong, is not a form of direct realism, since according to him phenomenalism implies that the direct or immediate objects of perception are the perceiver's own conscious experiences (sense-impressions, sense-data).[12]

But a phenomenalist does not have to say that the immediate objects of awareness in perception are conscious experiences. Rather, according to a phenomenalist, the immediate objects of perception are physical objects and events. Conscious mental states are not something that we typically perceive: they are something that we typically *have,* without perceiving them. We *perceive* physical things by *having* conscious experiences—experiences that fit into the totality of experiences in a certain way. (Maybe I can be said to perceive my own mental states when I introspect them, but introspection is not required for the perception of an ordinary physical thing, like a stone.) This is the view of perception taken by Mill, who therefore counts as a direct realist, in Armstrong's sense.

7.5 Conclusion

The goal of this chapter has been to show that phenomenalism is at least as defensible as any other idealist metaphysic. Phenomenalism makes the order and coherence of conscious experience no more mysterious than any other brand of idealism—or if it does, it makes other things correspondingly less mysterious. Phenomenalism does not imply that we are trapped behind a veil of ideas, each of us perceiving nothing but his own conscious experiences. And phenomenalism has the advantage over other narrow idealisms of avoiding commitment to the claim that physical things depend for their reality on the actual existence of something mental.

Of course, phenomenalism still faces all the objections that pertain to idealism in general. But given that a phenomenalist theory can achieve just as much as any other idealist theory, and with fewer resources, it seems reasonable to prefer phenomenalism to other idealist metaphysics, now that we know that

[12] Armstrong (1961), xi.

phenomenalism can handle all of the objections that target those features of the phenomenalist position that distinguish it from its idealistic competitors.

We already knew that if we were going to banish consciousness from spacetime, we would have to be idealists; the moral of this chapter is that if we are going to be idealists, we ought to be phenomenalists.

8

Radical phenomenalism

> In a word, psychologic time is given to us and must create scientific and physical time.
>
> Henri Poincaré, *The Measure of Time*

8.1 Introduction

Our goal is to find a defensible metaphysics that does not locate conscious experience in time or space. We have settled on phenomenalism as the best prospect, but it remains for us (1) to formulate phenomenalism in an explicitly atemporal way, and (2) to defend phenomenalism from the major objections that face any idealist metaphysic.

In this chapter, we take on the first of these tasks.

But before we get to that, I should point out that if phenomenological atemporalism gives its proponents a reason to favor phenomenalism, likewise phenomenalism (and idealism generally) gives its proponents a reason to favor phenomenological atemporalism. This is because in order to reduce the physical facts to phenomenological facts, we have to reduce facts about how things are arranged in physical time to phenomenological facts, which is possible only if phenomenological entities are not in physical time.[1]

[1] "If the Phenomenalist is to proceed in an honest manner, he must build up the conception of public time without appealing to temporal relations holding between the experiences of different minds. For they are part of what has to be built up." Armstrong (1961), 66.

Perhaps the closest that any historical idealist has come to appreciating this fact is Leibniz; although see also Ayer (1946–7), 181–2. One of the most intriguing ideas to emerge from Leibniz's metaphysical writings is his suggestion that time and its contents admit of idealist reduction no less than space and its. Both time and space are "phenomena, like the rainbow." Leibniz (1714/1989b), 307. [See also Leibniz (1712/1989), 199.] This is connected with Leibniz's cryptic remark that "time is the order of inconsistent possibilities." Leibniz (1703/1989), 178. The idea here seems to be that time can be understood as the possibility for a thing to differ from itself—an understanding at which Leibniz may have arrived by starting with the precept that time is equivalent to the possibility of change, and analyzing change as the possession of incompatible properties by a single thing. Since the only things that exist, on Leibniz's view, are monads, we are left with a conception of time as the possibility of

The need to idealize time might have seemed less pressing to Berkeley, Kant, and Mill than to a modern-day idealist. These historical idealists might have settled for a reduction of spatial facts, leaving the temporal aspects of physical reality unreduced. But a modern conception of spacetime rules out such half measures; as we have seen, on the modern view, time and space are too intimately intertwined to allow an analysis of one independently of the other. So a modern-day idealist really has no choice but to idealize facts about time as well as space—facts about how things are arranged in spacetime.[2]

The goal of this chapter is to provide a phenomenalist metaphysics that does just that. In Section 8.2, we undertake a critical examination of Mill's phenomenalism, focusing on Mill's development of the notion of phenomenological potential. We consider a variety of objections to Mill's account. While none of these seems clearly insurmountable, it must be admitted that Mill himself does not do a very good job of addressing them. Seeing where Mill's account is incomplete or inadequate will enable us to improve on his efforts in Section 8.3, where I develop the central theory of this book.

8.2 Lessons from Mill

All idealist theories are at odds with the common-sense view that conscious experience arises from physical processes and events. But there is one respect in which Leibniz, Berkeley, and Kant do not depart from common sense. They all think of consciousness as having some *source,* albeit not a physical source, and they all conceive of phenomenological potential as somehow grounded in this source. For Leibniz, it is grounded in the monads; for Berkeley, it is grounded in God; for Kant, it is grounded in the things-in-themselves. This is in sharp contrast to Mill, who conceives of phenomenological potential as the bottom-most level of reality.

monadic self-difference; see Leibniz (1685/1999), 629, Leibniz (1680/1999), 398, and, for further references and discussion, Futch (2008), 131–5.

Despite his tantalizing remarks on the ideality of time, however, Leibniz tends to describe monads and their experiences in flatly temporal terms. For example, he writes that "since every present state of a simple substance [i.e. monad] is a natural consequence of its preceding state, the present is pregnant with the future." Leibniz (1714/1989a), 216. Charles Hartshorne comments on the uncertain role of time in Leibniz's metaphysics in Hartshorne (1946), 420–1.

The most ambitious idealism of the past hundred years is the phenomenalism of Carnap's *Aufbau;* however, Carnap fails to phenomenalize time, since he takes as his primitive relation "recollection of similarity," which is defined in explicitly temporal terms; see Carnap (1967), 127–8.

[2] Could an idealist hold that experience occurs in spacetime? Such a view is no longer idealism, since it cannot aspire to reduce even the *spatial* aspects of the physical world to something phenomenological. Rather, the view is some kind of panpsychism, for criticism of which see Chalmers (2012).

There is an analogy here with certain developments in early nineteenth-century geometry. Prior to that time, mathematicians investigating the properties of curves of a given dimension always thought of the curves as embedded in some higher dimension. For example, spherical geometry was understood as an investigation of the properties of the two-dimensional surface of a three-dimensional Euclidean sphere, rather than a curved two-dimensional space. The idea of a "curved space" made no literal sense—or so it seemed.

It was Gauss who showed, with his famous *Theorema Egregium,* how to understand the curvature of a two-dimensional surface as an intrinsic property of the surface, without thinking of the surface as a boundary of some underlying three-dimensional structure. This eventually led to the realization that one could speak quite literally of a curved space—even a curved space of three or more dimensions—without thinking of that space as embedded in a "real" space of higher dimension.[3]

The central challenge for phenomenalism is to find a metaphysical analog of the *Theorema Egregium*—some way to conceive of phenomenological potential in intrinsically experiential terms, rather than in terms of the relationship of experience to some underlying physical, monadic, or noumenal structure. The phenomenalist seeks a notion of *intrinsic phenomenological potential,* akin to Gauss's notion of intrinsic geometric curvature.

As we have seen, Mill understands this intrinsic potential in terms of a certain kind of counterfactual dependence. For Mill, the obtainment of a given physical state of affairs reduces to the truth of a conjunction of sensation-conditionals:

> If there were to occur an appearance of its being the case that p_1, there would occur an appearance of its being the case that q_1; and if there were to occur an appearance of its being the case that p_2, there would occur an appearance of its being the case that q_2; and . . . if there were to occur an appearance of its being the case that p_n, there would occur an appearance of its being the case that q_n

for some presumably large (Mill suggests infinitely large) n.[4] A conjunction of true sensation-conditionals that intuitively entails the obtainment of some physical state of affairs is what Mill calls a Permanent Possibility of Sensation. The physical world is, according to Mill, nothing but a (vast, possibly infinite) collection of Permanent Possibilities of Sensation.

Mill's account is the natural point of departure for any attempt to revitalize phenomenalism. But his account is unclear on certain crucial points, and some

[3] The classical source here is Gauss (1902).

[4] For the suggestion that n might be infinite, see Mill (1865/1979), 179.

of the things that Mill says in developing it do not sit well with the phenomenalist world-view. To be fair, Mill develops his phenomenalism in the course of a lengthy critique of the philosophy of the now-forgotten Sir William Hamilton. Bearing this context in mind, we should not expect Mill's discussion to display the rigor and attention to detail that would characterize a more focused development of the theory. But sometimes Mill's carelessness leads him into confusions too serious to ignore.

For example, in explaining how a phenomenalist can account for the existence of unperceived things, Mill writes:

> I believe that Calcutta exists, though I do not perceive it, and that it would still exist if every percipient inhabitant were suddenly to leave the place, or be struck dead. But when I analyse the belief, all I find in it is, that were these events to take place, the Permanent Possibility of Sensation which I call Calcutta would still remain; that if I were suddenly transported to the banks of the Hoogly, I should still have the sensations which, if now present, would lead me to affirm that Calcutta exists here and now.[5]

In other words, Calcutta can exist unperceived, inasmuch as it is possible for there to be a situation in which (1) there are no perceptions of Calcutta, even though (2) it is true that if someone were standing on the banks of the Hoogly, he would have experiences as of a city with certain features (features distinctive of Calcutta).

Mill tries to account for intermittently perceived things in the same way:

> I see a piece of white paper on a table. I go into another room. If the phænomenon always followed me, or if, when it did not follow me, I believed it to disappear *è rerum naturâ*, I should not believe it to be an external object. I should consider it as a phantom—a mere affection of my senses: I should not believe that there had been any Body there. But, though I have ceased to see it, I am persuaded that the paper is still there. I no longer have the sensations which it gave me; but I believe that when I again place myself in the circumstances in which I had those sensations, that is, when I go again into the room, I shall again have them; and further, that there has been no intervening moment at which this would not have been the case.[6]

Mill's accounts of unperceived and intermittently perceived things are inadequate because they invoke counterfactuals with unreduced antecedents. Calcutta can exist unperceived because even in a scenario in which there are no perceptions of it, it remains the case that *if I were standing on the banks of the Hoogly*, I would have such-and-such experiences; the paper is intermittently perceived because between two times at which I perceive it, there are times when I do not perceive it, but would perceive it *if I were in the room*. But what is it, to stand on the banks of

[5] Mill (1865/1979), 184. [6] Mill (1865/1979), 179.

the Hoogly, or to be in the room? These are physical states of affairs in their own right, standing as much in need of phenomenalistic reduction as any others.[7]

In order to carry out his program successfully, Mill must not utilize such unreduced antecedents. He must confine himself to purely phenomenological counterfactuals, that is, to proper sensation-conditionals.

Let us focus on the case of unperceived objects. What Mill should do is to equate these with unrealized Permanent Possibilities of Sensation, understood as conjunctions of true sensation-conditionals having false antecedents.

For example, suppose that Calcutta has been abandoned, and now exists entirely unperceived. A Millian analysis of this state of affairs would go something like this: "If one were to have experiences as of standing by the side of a river with such-and-such features and facing in a certain direction, one would have experiences as of a city with so-and-so features"—where "such-and-such features" are those that we think of as characterizing the Hoogly, and "so-and-so features" those that we think of as characterizing Calcutta.

Apart from its incompleteness (pending proper substitutions for "such-and-such" and "so-and-so"), this analysis suffers from a defect noted by Smart:

> Consider the statement 'The walls of the bathroom are blue.' At a first shot the phenomenalist might analyse this as 'If I were in the bathroom I would be having blue sense impressions.' But what does 'in the bathroom' mean here? These words would themselves have to be analysed in terms of actual and possible sense impressions. Perhaps we might replace the reference to the bathroom by the phrase 'if I had such and such sense impressions' (e.g. of white bath, red tiles, green mat). Our original sentence would then become, say, 'if I had sense impressions as of white bath, red tiles, green mat, then I would have sense impressions as of blue walls.' Even so, our analysis is incomplete. I might have sense impressions as of white bath, red tiles, green mat in a similar bathroom which had yellow walls.[8]

Smart begins by pointing out that a phenomenalist cannot appeal to unreduced antecedents (as Mill in fact does in the passages quoted above), and then considers the natural-seeming solution of replacing these antecedents with antecedents of purely phenomenological import. The problem with this, however—as Smart observes—is that it leaves us with a sensation-conditional that fails to capture what we actually mean when we say (for instance) that the walls of the bathroom are blue.

[7] H.H. Price makes the same point against Hume: see Price (1940), 30–2. Kenneth Winkler suggests that something like Mill's account of unperceived things might solve Berkeley's problem of unperceived things, apparently unaware of the difficulties facing Mill's account; see Berkeley (1710/1982), xxxiii–xxxiv.

[8] Smart (1963), 20; see also Valberg (1992), 161–2.

The same considerations apply to Mill's own examples. Suppose that on some distant planet much like Earth, there is a river, the Eelgoo, that looks, sounds, feels, smells, and in every other observable respect resembles the Hoogly. To the east of the Hoogly lies Calcutta; to the east of the Eelgoo lies a dense jungle.

Now, in this situation—which is certainly a possible one—it is *false* that if one were to have experiences as of standing by the side of a river with such-and-such features (those that characterize the Hoogly) and facing East, one would have experiences as of a city with so-and-so features (those that characterize Calcutta). This sensation-conditional is false because, in the nature of the case described, one might have experiences as of a jungle rather than a city, were one to have experiences as of standing on the banks of a river with the features that characterize the Hoogly.

Of course, we have no reason to think that there actually is another river just like the Hoogly. But that is cold comfort to Mill. For (1) Mill's analysis makes the truth of "Calcutta exists" depend on the falsity of "The Eelgoo exists," whereas there is in fact no such dependency, and (2) there are plenty of actually existing environments that are perceptually indistinguishable from one another (Smart gives the example of two places near the South Pole).[9]

A Millian could try to get around these problems by focusing on Calcutta itself, rather than its environs. He might analyze "Calcutta exists" as: *If one were to have experiences as of a city with such-and-such qualities, one would have experiences as of a city with such-and-such other qualities,* where the first and second sets of qualities are ones that we think of as partly characterizing Calcutta.

But the problem recurs. On some distant planet there could be a city, Uttaklak, that is observationally indistinguishable from Calcutta. (Perhaps Uttaklak stands on the banks of the Eelgoo, where a dense jungle once flourished . . .) The above-italicized sensation-conditional would be true in this case, even if Calcutta did not exist (indeed, even if the Earth itself did not exist). This shows that the conditional is not equivalent to the proposition that Calcutta exists.

To overcome this difficulty, a Millian may shift the focus of his theory from individual experiences to whole streams of consciousness.[10]

Thus, instead of analyzing "Calcutta exists" in terms of the sort of experiences one would have if one were to have experiences of another sort, Mill might analyze it in terms of the features that one sort of stream of consciousness would have,

[9] Smart (1963), 21.

[10] Such a shift has some precedent in Mill's own presentation of his theory; see, for example, Mill (1865/1979), 190, where he characterizes the mind as "a series of feelings, or, as it has been called, a thread of consciousness, however supplemented by believed Possibilities of consciousness which are not, though they might be, realized."

given that it had certain other features. Schematically, the analysis might look something like this:

> Calcutta exists just in case it is true that if there were an n-element stream of consciousness (a stream comprising n sequential phenomenal states) whose first n-minus-1 elements had phenomenal characters $C_1, C_2, \ldots C_{n-1}$, then the nth element of that stream would be an appearance of a city having such-and-such qualities,

where "such-and-such qualities" are again qualities that we think of as characterizing Calcutta. This shift to streams of consciousness lets us take advantage of the fact that traveling wide-awake to Calcutta from a given location—say, Detroit—would involve a different stream of consciousness from the stream that traveling from Detroit to Uttaklak would involve. (The voyage to Uttaklak, unlike the voyage to Calcutta, would involve experiences of interstellar travel.)

At this point, we have moved some distance from Mill's original theory. In Section 8.3, we'll move even farther in the same direction, in the course of developing the phenomenalist position more fully than Mill himself ever attempted to develop it. But first, we should take note of another problem that Mill fails to address.

Conscious experiences tend to be accompanied by experiences as of sentient bodies or their effects. When I pull up the anchor, I feel the tension in the anchor line, but I also feel the line pressing against my palm. When I walk through the woods, I hear birds, but the sounds they make differ from what they would have been, had I been absent. And no matter what I am doing, I generally have proprioceptive experiences of my own body (its temperature, and the position of its limbs) and experiences as of my body's distance and orientation relative to various objects in my vicinity.

Now, according to Mill, a piece of paper in an unoccupied room exists even though no one perceives it, just because it is true that if there were experiences as of entering a certain room, they would be accompanied by experiences as of a certain piece of paper. But it is also true that if there were experiences as of entering the room, they would be accompanied by experiences as of a sentient body (the body of whoever, or whatever, was having the experiences as of entering), or as of the effects of such a body, or as of the spatial orientation of such a body relative to objects in its vicinity. So it seems that Mill must say that the room contains not only a piece of paper, but a sentient body. But the room does not contain any such body, and even if it did, this would not be a logically necessary requirement for its containing a piece of paper.

In fact, there is a way for Mill to get around this "extra-bodies" problem. The key is to recognize that an appearance of embodiment need not entail the truth of a sensation-conditional that guarantees the existence of a corresponding body.

How this is so will emerge in the discussion that follows, where we leave Mill behind to develop our own version of phenomenalism.

8.3 The central theory

Suppose that mankind acquires the ability to travel unlimited distances through space. Setting out from Earth in all directions, astronauts explore the Solar System, the Milky Way galaxy, and the universe beyond. Suppose furthermore that mankind acquires the ability to travel through time; to mitigate qualms about the intelligibility of this supposition, we can assume that time travelers are able to perceive but not influence the events they observe while time-traveling.[11] Suppose, finally, that medical science advances to the point where it is possible to extend people's lives indefinitely, so that human beings are effectively immortal.

Consider the life of one of these futuristic explorers, as he travels through time and space. He begins, let us suppose, by exploring nearby regions of space and nearby eras of time: star systems within a thousand light years of the Sun, and events within a thousand years of the present. Once he has observed and documented everything within this region of spacetime, he proceeds to explore remoter reaches of the universe: star systems out to a distance of a million light years, and eras within a million years of the present. And so he continues, exploring the cosmos in an ever-expanding sphere of inquiry, forever, or until he determines that there is no spacetime left to explore.

Among the things that the explorer hopes to discover are extraterrestrial life forms. Of course, whether he discovers any such life depends on whether living things ever actually evolve anywhere besides here on Earth. But what we can say for certain is that *if* there is extraterrestrial life, then the explorer is bound to perceive extraterrestrial life at some point of his exploration. Of course, it is also true that if the explorer perceives extraterrestrial life in the course of his exploration, then there is indeed extraterrestrial life at some time and place.

Roughly, the phenomenalist theory that I want to propose says that a physical thing x exists if, and only if, there would be experiences interpretable as perceptions of x, if someone were to have the kind of experiences that the hypothetical spacetime explorer has; that is, (roughly) experiences as of exploring all of time and space.

Now, for us, there is no such thing as an experience as of exploring all of time and space. Indeed, for us, there is no such thing as an experience as of everything that happens in a café over the course of an afternoon. Human experiences are

[11] See Casati and Varzi (2001).

typically as of the contents of brief intervals of time and smallish regions of space. So, if we think of the spacetime explorer described above as having more or less human experiences, we cannot think of all his experiences as parts of a single, vastly complex experience. At most, we might think of them as belonging to a single, vastly complex stream of consciousness.

Let us call the stream of consciousness, or the set of streams of consciousness, that characterize the explorer's mental life an *experiential voyage*.

I have just mooted the concept of an experiential voyage by reference to a hypothetical sentience that travels throughout time and space, making conscious observations as he goes along. Now I want to characterize an experiential voyage in purely phenomenological terms, without implicit reference to an independently-given physical world. (We can think of this as the first stage of the project of developing a notion of intrinsic phenomenological potential, akin to Gauss's intrinsic geometric curvature.)

In Chapter 6, I introduced the idea of an interpretation of a totality of experiences, pointing out that a given experience might be best interpreted as a perception of one thing when interpreted as part of one set of experiences, and best interpreted as a perception of another thing (or of nothing) when interpreted as part of a broader set of experiences. Let us say that a given set of experiences "evokes" some physical object or event, *x*, if and only if the set is interpretable as including a perception of *x*.

Suppose one set of experiences has another set of experiences as a subset. Suppose furthermore that the superset evokes the observable contents of a spherical region of spacetime, and the subset evokes, within the context of the superset, the observable contents of a region of spacetime that lies at the center of the spherical region whose contents the superset evokes.[12] Then I say that the superset is a *supergroup* of the subset, and the subset a *subgroup* of the supergroup:

Definition

Given two sets of experiences, S, S', we say that S' is a supergroup of S if and only if (1) S is a subset of S', (2) S' evokes the contents of a spherical region of spacetime (i.e. S' is interpretable as including perceptions of the observable contents of a spherical region of spacetime), and (3) interpreted as part of S', S evokes the contents of a region of spacetime that lies at the center of the region whose contents S' evokes.

An example of two sets of experiences standing in the subgroup/supergroup relation is the set S of all the experiences that the spacetime explorer has in the first phase of his exploration (out to a thousand years and a thousand light years),

[12] Strictly speaking, I should speak of a pseudo-spherical or hyperboloidal region, since we are talking about a region of spacetime. I leave this understood for the remainder of the discussion.

and the set S' of experiences that includes all the experiences in S, plus all the experiences that the explorer has in the course of the second phase of his exploration; here, S is the subgroup, and S' the supergroup. In turn, S' is a subgroup of the supergroup S'' that includes all the experiences in S', in addition to all the experiences the explorer has in the third phase of his exploration; etc.

We can consider an infinite series of groups of experiences, each adjacent pair in the series standing in the subgroup/supergroup relation. Call any such series a *phenomenological group series.*

So far we have not introduced any quantity, "phenomenological volume of x," but only a relation of "x evoking a smaller region of spacetime than y, in the context of y." Let us now define the *phenomenological volume* of a group of experiences as the smallest volume such that the physical events that the group evokes are evoked as occurring within a spherical region having that volume.

Finally, suppose that a phenomenological group series $\langle G_1, G_2, G_3, \ldots G_n \rangle$ satisfies the following condition: as n approaches infinity, so does the phenomenological volume of G_n. Call such a series a *phenomenological expansion*. Reading "$V(x)$" as "the phenomenological volume of x," we have:

Definition
A phenomenological expansion is a phenomenological group series $\langle G_1, G_2, G_3, \ldots G_n \rangle$ such that $\lim_{n \to \infty} V(G_n) = \infty$.

Using the notion of a phenomenological expansion, we might characterize an experiential voyage as follows: an experiential voyage is a collection of experiences that includes all and only the experiences that belong to the members of a phenomenological expansion. More precisely: an experiential voyage is a collection of experiences C such that there is a phenomenological expansion E such that for all x, x belongs to C if, and only if, there is a member of E to which x belongs.

The notion of a phenomenological expansion does a good job of capturing the idea of an experiential voyage—good, but not perfect. Certainly, it allows us to understand experiential voyages without implicit reference to an independently-given world of things in time and space. The problem is that characterizing an experiential voyage in terms of phenomenological expansion makes potentially controversial assumptions about the topology of spacetime.

The easiest way to see this is to reflect that if spacetime has a spherical geometry, in which every straight line can be extended to form a closed curve, then the volume of spacetime is finite. In that case, successive phases of the spacetime explorer's odyssey will not cover ever-larger regions of spacetime, but will eventually reach some limit, beyond which further exploration will reveal nothing new. If we define experiential voyages in terms of phenomenological expansions, as above, we will be implicitly ruling out the possibility that spacetime is finite in this way.

As it happens, there seems to be little reason (at present anyhow) to believe that the actual universe is characterized by closed world lines. Still, it would be inappropriate to prejudge in our metaphysics what is ultimately an empirical question. Furthermore, even if actual spacetime does not have a temporally closed structure, it is a known fact that a spacetime having such a structure is *possible*, and a phenomenalist ought to be able to say what it would be for there to exist things in a physical universe having that kind of structure. Since a characterization of experiential voyages in terms of phenomenological expansions would prevent us from doing this, we need to find some other purely phenomenological understanding of experiential voyages.[13]

Define a *linear region* of spacetime as a region of spacetime that consists of a straight line segment of spacetime together with every point of spacetime lying within some finite distance of that line segment. (For an image, you can think of this as a finite tube of spacetime.)

Suppose that G' is a set of experiences, and G a subset of G'.

Suppose furthermore that G, taken in the context of G', evokes the perceptible contents of linear regions R_1, R_2, R_3 . . . such that (1) there is a place at which all these linear regions intersect, and (2) for every angle, θ, there is a pair of linear regions (among R_1, R_2, R_3 . . .) whose angle of intersection is θ. Intuitively, this means that G evokes the perceptible contents of a finite region of spacetime consisting of an infinite cluster of overlapping linear regions of spacetime. (For an image, you can think of this as the region carved out by a cylinder that rotates around its center through all possible angles.)

Finally, suppose that G' evokes the contents of linear regions R'_1, R'_2, R'_3 . . . such that R'_1 subsumes R_1, R'_2 subsumes R_2, R'_3 subsumes R_3, etc. (A linear region X subsumes a linear region Y just in case the segment that defines Y is an internal sub-segment of the segment that defines X, as the segment of the number-line that goes from 2 to 4 is an internal sub-segment of the segment that goes from 1 to 5.)

Given these assumptions, we call G' a *linear extension* of G.

Now consider a series of sets of experiences, such that each set in the series (besides the first) is a linear extension of the preceding set in the series. Call such a series a *phenomenological extension series.*

There are two kinds of extension series.

First, an extension series Σ might contain an element that evokes everything that any subsequent element of Σ evokes. In other words, Σ might contain an

[13] The most famous example of a spacetime characterized by closed timelike curves is the Gödel solution to the Einstein field equations; see Gödel (1949), and for discussion, see Earman (1995), 160–202.

nth element, G_n, such that for all x, and all m greater than n, element G_m of Σ evokes x only if G_n evokes x. Call a phenomenological extension series that has this property a *bounded* series, and call the first element of a bounded series that plays the role of G_n the *bounding element* of the series.

Intuitively, a bounded phenomenological extension series is the kind of series that the cumulative totalities of the spacetime explorer's experiences constitute, if spacetime has a spherical geometry in which every straight line can be extended to meet itself, forming a closed curve in spacetime (see Fig. 8.1a). In that case, Σ begins with a sequence of groups of experiences, each of which evokes a larger region of spacetime than the group preceding it, until we get to a group $G^\star$ that corresponds to the point when the explorer's investigation of spacetime leads him back to the point from which he began. Any further exploration will yield "redundant" groups of experiences that evoke no physical entity that $G^\star$ doesn't already evoke. (Here, $G^\star$ is the bounding element of the series.)

Alternatively, an extension series might be unbounded. Intuitively, an unbounded extension series is the kind of series that the cumulative totalities of the explorer's experiences constitute, if spacetime has a geometry in which not every straight line can be extended to meet itself. One example of such a geometry is flat Minkowski spacetime, analogous to Euclidean space, in which no straight line meets itself (see Fig. 8.1b); another is a toroidal spacetime in which some, but not all, straight lines close in on themselves; yet another is the gently curved spacetime that appears to characterize actual physical reality. For a spacetime with this kind of geometry, there is no guarantee of a corresponding phenomenological extension series that contains a bounding element (a totality of experiences that renders subsequent elements interpretively redundant, in the manner of $G^\star$).

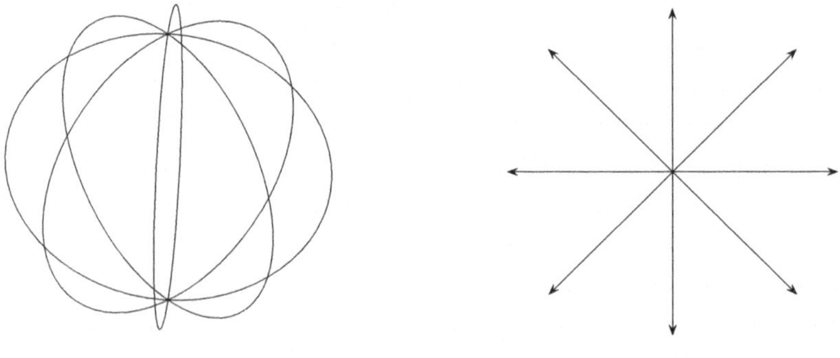

(a) spherical (b) flat

Figure 8.1 Straight lines in (a) spherical and (b) flat space.

With these definitions in place, we can now characterize an experiential voyage as a collection of experiences that includes all and only the experiences that belong to the members of a phenomenological extension series. More precisely: an experiential voyage is a collection of experiences C such that there is a phenomenological extension series Σ such that for all x, x belongs to C if, and only if, there is a member of Σ to which x belongs.

We have defined the concept of an experiential voyage in a purely phenomenological way. Now we must use the concept to give a phenomenalist analysis of spacetime and its contents.

Let us begin by introducing *voyage-conditionals,* analogous to Mill's sensation-conditionals. A voyage-conditional is a proposition of the form, "If there were an experiential voyage with phenomenological feature F, that voyage would also have phenomenological feature G," where "F" and "G" are understood to refer to different phenomenological features. Examples of phenomenological features are features of the form, *including an experience with such-and-such a phenomenal character,* and: *including a stream of consciousness (or a portion of a stream of conscious) comprising such-and-such overlapping experiences.* An example of a voyage-conditional is: If there were an experiential voyage that included a stream of experiences as of traveling to the far side of the Moon, that voyage would include experiences as of craters.

In terms of voyage-conditionals, we can already state a global version of phenomenalism. Call it:

Global radical phenomenalism
The physical facts (i.e. facts about spacetime and its contents) logically supervene on the facts expressed by true voyage-conditionals.

In other words, how things are physically is a logical consequence of which voyage-conditionals are true.

The global thesis captures the basic phenomenalist idea that physical reality consists of whatever we would perceive, or could reasonably be taken to perceive, were we to have experiences as of exploring all of time and space. Global radical phenomenalism is a clear descendant of Mill's phenomenalism. The main differences between global phenomenalism (as stated above) and Mill's phenomenalism are, first, that global phenomenalism is explicitly holistic, invoking voyage-conditionals instead of sensation-conditionals, and, second, that global phenomenalism idealizes facts about time and the temporal, in addition to facts about space and the spatial. As for the first difference, we saw in Section 8.2 that a Millian ultimately has to shift from talk of sensations to talk of streams, in order to deal with Smart's objection (the bathroom, Uttaklak and the Eelgoo, etc.); so in this respect, global phenomenalism is a natural development of Mill's position. As for the second difference, Mill refrained from phenomenalizing time only because he thought that an atemporal conception of experience was unintelligible. We have seen our way past this prejudice, allowing us to formulate a truly global phenomenalism in terms of voyage-conditionals pertaining to phenomenological structures (streams of consciousness, or collections of streams) that do not exist in time.

Global phenomenalism expresses the core idea of radical phenomenalism, but it is, in a way, severely limited. Global phenomenalism is a phenomenalist reduction of physical reality *as a whole*, but it does not imply an analysis of any individual *part* or aspect of physical reality. For example, it does not imply an analysis of the fact that there is a pineapple on the table. To achieve this, we need to formulate a *local* phenomenalism by which we can analyze physical reality piece by piece, rather than as a vast undifferentiated chunk.

As a first step toward such an analysis, let's define a special class of experiential voyage-conditionals, which we may call the *complete voyage-conditionals*. A complete voyage-conditional has the form,

> If there were an experiential voyage v with phenomenological features $A_1, A_2, A_3, \ldots$ then v would also have phenomenological features $B_1, B_2, B_3, \ldots$ and no phenomenological features besides the A-features and the B-features,

where the A-features are understood to be different from the B-features.

A complete voyage-conditional states, in effect, that if there were an experiential voyage with certain qualities, its phenomenological features would be exhausted by those qualities together with certain other qualities. It is a "complete" voyage-conditional, in the sense that it describes a counterfactual experiential voyage by reference to *all* of that voyage's phenomenological features (or would-be phenomenological features).

There are an infinite number of logically possible experiential voyages, few of which are relevant to a phenomenalist analysis of the physical. Our interest lies with a subset of the logically possible voyages, which we can call the *potential voyages*. A potential experiential voyage v is a logically possible voyage that would tend to confirm some true complete voyage-conditional, if v actually existed. More precisely, a logically possible experiential voyage v is a *potential* voyage if, and only if, v satisfies the following conditions: (1) for some set of phenomenological features A and some distinct set of phenomenological features B, v has (at any possible world in which it exists) the features in A, the features in B, and no phenomenological features besides; and (2) if there actually were a voyage with the A-features, it would have the B-features, and no features besides the A-features and B-features.

Intuitively, a potential experiential voyage is one that is *poised for realization*. It is a possibility that lies close to the modal surface, so to speak, in contrast to mere logical possibilities whose actualization would not tend to confirm any counterfactual conditional that is both complete and true. There is a logically possible stream of consciousness involving experiences as of traveling to the Moon to discover that it is made of Limburger cheese. The existence of this logical possibility has no bearing on the existence or nature of the Moon, since there is no sense in which the possibility is poised for realization. To the Limburger possibility there corresponds no truth to the effect that if there were experiences as of traveling to the Moon, then there would be experiences as of discovering that the Moon is made of cheese. This is in contrast to the possibility of a stream of consciousness involving experiences as of traveling to the far side of the Moon to discover that it is covered with craters. To this possibility there *does* correspond a relevant conditional truth—a truth to the effect that if there were experiences as of traveling to the far side of the Moon, there would be experiences as of discovering that the far side is covered with craters.

We can use the notion of a potential experiential voyage to state a local version of radical phenomenalism. The simplest such analysis is this:

For all x, x exists in spacetime if, and only if, the totality of all potential experiential voyages evokes x.

Here we extend the notion of "evoking" so that it applies not just to totalities of experiences, but also to totalities of streams of consciousness (or totalities of sets of streams of consciousness). A set of experiential voyages evokes x just in case the voyages in the set are collectively interpretable as including perceptions of x, where interpretation is understood as explained in Chapter 6.

Why analyze the existence of a physical thing in terms of *all* potential experiential voyages? Why isn't it enough for the existence of some physical thing, like an apple, that *some* potential voyage includes experiences interpretable as perceptions of that thing?

The reason for saying "all" instead of "some" is that there might be potential voyages that, when considered on their own, evoke various physical things, but that, when considered together with all potential voyages, do not evoke those things. For example, we cannot rule out the possibility that there is a potential experiential voyage corresponding to the experiences that an insane spacetime explorer might have. Taken on its own, an abnormal voyage like this might include experiences interpretable as perceptions of fairies, dragons, etc. The potential for such a voyage—if there is such a potential—does not compel us to posit the existence of fairies. Only if the experiences remain interpretable as perceptions of fairies when we interpret them in the context of *all* potential voyages should we conclude that there are fairies. But if, as we have every reason to believe, the experiences would not remain so interpretable in the broader context, we should not take their interpretability as perceptions of fairies in the narrower context as a basis for positing fairies.[14]

Now, it is possible, and even likely, that there are an infinite number of potential experiential voyages. One might complain that this renders the proposed analysis obscure, on the grounds that we have no firm grasp on what it means to speak of interpreting an infinite number of experiences (or an infinite number of streams of consciousness, or of sets of streams). To mitigate qualms relating to the possible infinitude of the totality of potential voyages, we can modify the foregoing analysis as follows. Letting "S_n" designate a random set of n potential experiential voyages, we have:

Local radical phenomenalism
For all x, x exists in spacetime if, and only if, the probability that S_n evokes x approaches 1, as n approaches infinity.

For a convenient gloss of this, we can say that physical x exists if, and only if, there are perceptions of x "in the phenomenological limit."[15]

The analysis above does not require it to be possible to interpret any infinite totality (of experiences, streams of consciousness, or whatever), and does not rely on the existence of anything like a "completed infinity" of the sort that certain philosophers and mathematicians find objectionable.[16]

Local radical phenomenalism analyzes individual physical entities in terms of random sets of potential experiential voyages. Why random sets? This is to ensure that there is no distorting bias in the potential voyages. Reverting to the futuristic explorers discussed above, we can imagine that for every million sane explorers, there is one insane explorer. If there are an infinite number of explorers, there will be an infinite number of insane explorers (as well as an infinite number of

[14] Another thing we cannot rule out is the possibility that potential voyages that include the experiences we actually have fail to evoke the sorts of things we take ourselves to perceive when those voyages are interpreted in the context of all potential voyages. We discuss this possibility in detail in Chapter 9, in connection with various putatively skeptical hypotheses (such that our experiences all arise from envatted brains).

[15] An equivalent statement of the local analysis is: for all x, x exists in spacetime if, and only if, $\lim_{n\to\infty} P(S_n \text{ evokes } x) = 1$.

[16] Especially pre-Cantorian philosophers and mathematicians influenced by Aristotle, such as Gauss: "I protest against the use of an infinite quantity as something completed, which is never allowed in mathematics. The infinite is but a *façon de parler*, by which we only mean to speak of a limit that certain ratios approach as close as you like, while others are permitted to grow without restriction." Gauss (1900), 216.

sane ones). If we consider successive sets of experiential voyages S_1, S_2, S_3, . . . , each set containing one more voyage than the preceding set, each of these sets might contain only voyages corresponding to the mental lives of insane voyagers. In that case, there would be no reason to expect the probability of the *n*th set in the series to evoke any physical *x*, as *n* approached infinity. By stipulating that the membership of successive sets of voyages are determined at random, we ensure that no selection bias will distort the analysis.

The limit version of the local analysis captures the basic idea behind phenomenalism just as well as the simpler, infinitary analysis. The basic idea behind phenomenalism is that physical reality consists of whatever we could reasonably be taken to perceive, if we were to have experiences as of exploring all of time and space. To make this idea vivid, I imagined a hypothetical explorer of time and space. Some physical entity *x* exists at some time and place if, and only if, the explorer perceives *x* at some stage of his exploration. But it is equally clear that if some physical thing *x* exists, then the probability that the explorer perceives *x* approaches 1 as the extent of his exploration of spacetime approaches infinity, and that if the probability that the explorer perceives *x* approaches 1 as his exploration extends to infinity, then *x* does indeed exist at some time and place. The limit version of local radical phenomenalism ("the probability that S_n evokes *x* approaches 1 as *n* approaches infinity") therefore does just as good a job of capturing the basic phenomenalist idea as the simpler, infinitary version ("the set of all potential voyages evokes *x*").

Both local and global phenomenalism, as stated above, have a major and obvious shortcoming. They both imply that the physical universe consists entirely of *perceptible* entities. This is a shortcoming because we have good reasons to believe that the physical universe includes many imperceptible entities, including the most basic physical entities. A satisfactory analysis of spacetime and its contents must recognize the existence, or at least the possibility, of such entities. A main goal of Chapter 9 is to find a way to extend the phenomenalist analysis to the imperceptible realm.

But before concluding the present chapter, let's see how radical phenomenalism solves the problem of extra-bodies that we raised against Mill's theory at the end of Section 8.2.

8.4 Solving the extra-bodies problem

According to Mill, the existence of an unperceived physical object or event is equivalent to the truth of a conjunction of true sensation-conditionals, each of which has a false antecedent. The radical phenomenalist account of unperceived things is equally straightforward. According to radical phenomenalism, a physical entity x exists unperceived if, and only if, (1) the probability that S_n evokes x approaches 1, as n approaches infinity, and (2) there are no actual perceptions of x. In terms of the infinitary analysis: x exists unperceived if, and only if, (1) the totality of all potential experiential voyages evokes x, and (2) there are no actual perceptions of x.

It was in the context of discussing Mill's account of unperceived things that we encountered the "extra-bodies" problem. Recall that for Mill, to say that his study contains a piece of paper that nobody perceives is to say that even though no one perceives the paper, there would be experiences as of the paper, if there were experiences as of being in Mill's study. The problem is that it is equally true that there would be experiences as of a sentient body, if there were experiences as of being in Mill's study, from which it appears to follow, by Mill's account, that there is a sentient body in Mill's study. More generally, Mill's account seems to imply that there are sentient bodies all over the place, since it seems that the truth of *any* sensation-conditional "were there appearances of x, there would be appearances of y" guarantees the truth of a corresponding sensation-conditional: "were there appearances of x, there would be appearances of a sentient body," viz., the body of someone having an experience as of x. Apart from the fact that this implies the existence of far more sentient bodies than actually exist, it also undermines Mill's account of unperceived things, since it means that every perceptible object is accompanied by some sentient being who perceives it.

The key to solving this problem is to insist on a holistic analysis of the physical.

Let us grant, for the sake of argument, that every experience is an *appearance* of (among other things) an embodying body. From this it does not follow, and it is not the case, that each experience is interpretable, in the phenomenological limit, as a *perception* of an embodying body. In particular, the potential but non-actual experiences corresponding to unperceived things are not interpretable as perceptions of embodying bodies. For while each of these experiences is an appearance of a body that embodies it, the existence of this body is not, so to speak, corroborated by further experiences in the totality of potential experiential voyages. A hypothetical experience answering to an unperceived thing may be an *appearance* of (among other things) an embodying body, but in the phenomenological limit, neither that experience nor any other experience is interpretable as a *perception* of such a body. The apparent body is, as it were, invisible to all but one mind—the mind whose experiences are appearances of embodiment by that body. It is a *merely* apparent body, with no presence in spacetime.

So, to the extent that a potential experience answering to an unperceived thing is an appearance of a body embodying that experience, the experience is deceptive. But that need not prevent the experience from making a positive contribution to the sum total of experiences (considered in the phenomenological limit), relative to the goal of constructing physical reality. Even if the potential experiences that correspond to unperceived objects are characterized by the phenomenology of embodiment, the totality of all appearances does not, in the limit, contain the third-person phenomenology that it would have to, in order for the appearances to bear interpretation as perceptions, rather than mere illusions, of embodying bodies. Thus, we are spared the need to posit embodying bodies where there are, in fact, none.

That said, embodiment might seem to pose a real challenge to the radical phenomenalist. From a radical phenomenalist standpoint, humans and other sentient animals appear to be a strange amalgam of the fundamental conscious stuff of the world, and the physical stuff that is, so to speak, immanent in the conscious stuff. I have conscious experiences, and I have a body. The challenge for the radical phenomenalist is to account for this fact, given that my body exists in time and space, and my experiences do not.

My experiences are not merely associated with a particular human body, but individually correspond to specific internal processes taking place in that body. In particular, there is a correlation between the experiences that make up my stream of consciousness, and certain physical events that occur in my brain.

Let us say that a physical process P *maps* a stream of consciousness S just in case the following conditions are satisfied: (1) for some number *n*, P comprises physical events $p_1, p_2, p_3, \ldots p_n$, and S comprises sequential experiences $e_1, e_2, e_3, \ldots e_n$; and (2) there is a one-to-one correspondence between the physical events and the experiences, such that for every pair of experiences in S, e, e', if e' is e's immediate successor in S, then the physical event in P to which e' corresponds depends on the physical event in P to which e corresponds.[17]

When a physical process maps a stream of consciousness, we can think of the process as a spatiotemporally extended picture of the non-spatial, non-temporal stream of consciousness.[18]

In order for a given physical process (or the body that contains it) to qualify as the embodier of a given stream of consciousness, it is necessary, but not sufficient, for that process to map that stream. It could be that two or more physical processes

[17] The dependencies among events in P are to be understood in the manner described in Chalmers (1996b).

[18] That there is such a mapping between neural and phenomenal states is clear from developments in neuroscience; that the mapping must be a very close one is the thrust of Chalmers (1996a), 247–75.

map a single stream of consciousness, but that only one of those processes embodies the stream. (But which one?) It could happen that there are two physical processes, each of which maps both of two streams of consciousness. (But which process embodies which stream?) There could be two streams of consciousness both mapped by the same physical process and that process alone. (But which stream, if either, does the process embody?)

To get a proper assignment of streams to bodies, we need to specify some further relation that a stream must bear to a body, in order for that body to count as embodying that stream. What relation is this?

The standard view is that it is a relation of causal dependence of the experiences constituting the stream upon the events constituting the physical process. This view is unfriendly to phenomenalism, which does not think of conscious experiences as depending on physical events, at least not causally. But an analogous view is available to the phenomenalist. Instead of saying that a physical process that maps a stream of consciousness embodies that stream just in case each physical event in the process causes the experience to which it corresponds (in the mapping), a phenomenalist can say that a process that maps a stream embodies the stream just in case each experience in the stream gives a high probability to its corresponding physical event.

Of course, the physical events and processes in question are all to be understood as phenomenological constructions. But apart from this, the phenomenalist alternative to the standard account of embodiment is not especially radical. It is hard to think of any philosophy of mind that does not see the existence of a conscious experience as giving a high probability to the occurrence of a corresponding physical event (or sequence of events). Certainly a physicalist must admit the existence of this kind of conditional probability, and it is hard to see why a dualist would deny that the contents of conscious minds have a strong statistical bearing on the states of corresponding brains (albeit not a bearing grounded in identity or strict entailment, as the physicalist supposes).

The account of embodiment that I have offered allows us to make sense of the idea that we are beings both conscious and embodied, in a way that is compatible with a phenomenalism that denies conscious experiences any location in time or space. At the same time, the account gives us a way to describe conscious experiences in temporal terms: our experiences are in time, in the derivative sense that they correspond to events that take place in time as part of the physical processes that embody the experiences. Thus, even though the radical phenomenalist denies that experiences exist in time in any literal sense, he can still speak of them as being in time metaphorically: my experiences are in time in a sense akin to that in which Julius Caesar is in Gibbon's *Decline and Fall of the*

Roman Empire. Qua conscious being, I do not exist in time or space, but I have a spatiotemporal counterpart to which I may refer in describing my experiences.[19]

8.5 Conclusion

Crudely stated, radical phenomenalism is the view that the Moon's existence comes to the fact that the more experiences there are, the more likely it is for there to be experiences interpretable as perceptions of the Moon—and likewise for all other physical phenomena. If someone asks why he should believe that the more experiences there are, the more likely it is for there to be experiences interpretable as perceptions of the Moon, the answer is that if he *shouldn't* believe it (or the cleaned-up version of it represented by local radical phenomenalism), then he shouldn't believe that the Moon exists either.

Like Leibniz's monadology, the position developed here is an explicitly holistic idealism that aspires to reduce not just facts about space and the spatial, but also facts about time and the temporal, to phenomenological facts. Like Mill's phenomenalism, radical phenomenalism understands potential for conscious experience as the really fundamental thing about the world, ungrounded in underlying minds or things-in-themselves. At a detailed level, there is little that is really new about the theory I have advanced. What is new is how the details fit together. Radical phenomenalism is a maximal construction out of minimal resources. There is no more ambitious metaphysic.

Radical phenomenalism has many attractions. But the theory faces several unanswered objections. One of these we have already mentioned: the objection that radical phenomenalism cannot account for physical phenomena that are not merely unperceived, but imperceptible. Another is that radical phenomenalism fails to distinguish between veridical and non-veridical experiences. Another is that no amount of phenomenological potential can fully determine any physical state of affairs. We address these objections in Chapter 9.

[19] While the account I have offered makes sense of embodiment in a phenomenalistically acceptable way, it does not imply that all experiences actually are embodied. And even if all experiences are embodied as a matter of fact, my account of embodiment does not imply that experiences logically *must* be embodied. The account is compatible with the idea that there could be, or even that there actually are, unembodied minds.

9

Idealism vindicated

If you tell people that you're defending idealism, no one will read your book.

Anonymous

9.1 Introduction

Limit phenomenalism is a kind of idealism, and idealism faces three major challenges that we have yet to tackle. They are:

The Problem of Unusual Origins
No amount of phenomenological potential is *sufficient* for the existence of ordinary physical things, since we can imagine scenarios in which the potential exists, but the physical things do not.

The Problem of Deceptive Appearances
Idealism cannot distinguish between veridical appearances, of the sort we have in the course of ordinary waking perception, and deceptive appearances, of the sort that characterize dreams, hallucinations, misperceptions, etc.

The Problem of Imperceptible Things
Idealism is incompatible with the existence of imperceptible objects and events in spacetime.

In this chapter, we consider, and attempt to solve, each of these problems.

9.2 Unusual origins

Among the more popular pieces of contemporary philosophical folklore are stories in which all of our experiences arise from interactions between some envatted brains and a powerful computer to which the brains are connected. We are to suppose that the computer monitors the brains, and stimulates them so that they have various conscious experiences—experiences indistinguishable from those that we humans actually have. The supercomputer has the power to cause any humanly possible experience; which experiences it actually causes (in the envatted brains) depends on the computer's software and the information it gets from monitoring the brains. In short, the supercomputer/brains set-up harbors all the potential for conscious experience that the actual world harbors.

Finally, let us suppose that outside the vat set-up there are no trees, mountains, galaxies, or any other ordinary physical things.

Call the proposition that this whole story is true the *Vat Hypothesis*.

Is the Vat Hypothesis compatible with the existence of mountains and lakes, fields and streams, thunderstorms, trees, and galaxies? On a first appraisal, most people are inclined to say that it is not. Yet, it appears that an idealist must say that the Vat Hypothesis entails the existence of all these things and more. After all, the Vat Hypothesis implies that there are things (or a thing) with the power to cause experiences just like those that arise, or have the potential to arise, in our world. It follows that whatever phenomenological potential exists in our world also exists in the world described by the Vat Hypothesis. If, as idealists contend, the physical facts reduce to facts about phenomenological potential, it would seem to follow that whatever physical facts actually obtain (i.e. obtain in our world—the actual world) also obtain in the situation that the Vat Hypothesis describes.

Is this a refutation of idealism? To find out, let's frame it as an argument, which I'll call the *Argument from Unusual Origins*

D1. Idealism implies that there are trees, lakes, galaxies, etc. in any state of affairs that has all of the phenomenological potential that characterizes the actual state of affairs.

D2. The state of affairs that the Vat Hypothesis describes has all of the phenomenological potential that characterizes the actual state of affairs.

D3. So, if idealism is true, then the Vat Hypothesis describes a state of affairs in which there are trees, lakes, galaxies, etc. (D1, D2)

D4. But there are no trees, lakes, galaxies, etc. in the state of affairs that the Vat Hypothesis describes.

D5. Therefore, idealism is false. (D3, D4)

The best way for an idealist to respond to this argument is by rejecting D1. Idealism does not imply that there are trees, lakes, etc. in every state of affairs that has all of the phenomenological potential that characterizes the actual state of affairs. At most, it implies that there are trees, lakes, etc. in every state of affairs that has all *and only* the phenomenological potential that characterizes the actual state of affairs. It implies this, assuming that there actually are trees, lakes, etc.—an assumption which I'll take for granted for the remainder of this discussion.

The situation that the Vat Hypothesis describes contains all of the phenomenological potential that characterizes our actual situation, but it also contains some phenomenological potential that our actual situation does not contain (or so we are assuming). The situation that the Vat Hypothesis describes is one in which there exist potentials for experiential voyages over and above actually existing potentials for experiential voyages. The totality of potential voyages that characterizes actuality is a subset of the totality that characterizes the vat scenario. The latter totality—call it the "vat-totality"—evokes a vat, some envatted brains, and a supercomputer. The actual totality does not.

Is the vat-totality interpretable as including perceptions of trees, lakes, etc.? The most natural thing to say is that it is not. If we focus on the potential experiential voyages associated with the envatted brains, the voyages we consider will indeed evoke trees, lakes, and the rest. But then we would be focusing only on a subset of all the potential voyages that characterize the scenario described by the Vat Hypothesis. When we take into account *all* of the potential voyages that characterize the vat scenario, we find that the experiences we previously interpreted as perceptions of trees and lakes are actually not best interpreted as perceptions of trees and lakes, but rather as perceptions of computational processes.

Now, one might dispute this. One might suggest that even in the context of all the potential voyages that characterize the vat scenario, there are experiences interpretable as perceptions of trees and lakes. It's just that the trees and lakes have an unexpected nature: a purely computational nature.

This is the line that David Chalmers takes. According to Chalmers, the Vat Hypothesis is entirely consistent with the proposition that there are, for example, trees. The Vat Hypothesis does not imply that trees do not exist; rather, according to Chalmers, it implies that trees have an unexpected nature—that the existence of a tree reduces to facts about the experience-causing powers of a powerful computer.[1]

The key idea behind Chalmers's proposal is that when we say things like "There is a tree," or "The lake is frozen," or "We live in a spiral galaxy," we do not mean to commit ourselves to any proposition about the ultimate origins of our experiences of trees, lakes, or galaxies. Commitment to the existence of trees does not carry with it commitment to any particular metaphysical scheme.

Support for this idea comes from the fact that historical shifts in basic ontology have not precipitated corresponding waves of external world skepticism. For example, the shift from a Four Elements ontology (according to which everything

[1] See Chalmers (2010a), 459–73, as well as Cargile (1997), 79–80.

we perceive fundamentally consists of earth, air, fire, and water) to a modern scientific ontology (according to which everything we perceive fundamentally consists of elementary fields and particles) did not lead people to adopt a skeptical or anti-realist attitude toward the physical universe.

In his discussion, Chalmers explores the Vat Hypothesis mainly from the perspective of the envatted brains. But what if there are conscious beings who perceive the vat, the brains, and the computer from the standpoint of outside observers, for example, from the standpoint of people who programmed the computer, put the brains into the vat, etc? Even if we suppose that there are no such beings, the Vat Hypothesis seems consistent with the *possibility* of such beings, and that is already enough to raise the question of what to make of the experiences that such beings would have, if they existed.

Here I am, the Vatmaster, watching the brains bob and the computer compute. According to Chalmers, the existence of this computer with its experience-causing powers is sufficient for the existence of what the brains call "trees," "lakes," and "galaxies." But I can plainly see that neither the computer, nor the brains, nor the vat, nor any region of the space I inhabit contains any trees, lakes, or galaxies. Doesn't this prove that Chalmers is wrong to suggest that there are trees in the situation that the Vat Hypothesis describes?[2]

If this line of reasoning seems persuasive, it is because we think that what the brains call "trees" are not really trees. But what is the basis for this?

The bats that flap around my house in the evening perceive the same trees I do, but by having conscious experiences utterly unlike any that I have, or can even imagine having. If I were suddenly to have the experiences that one of the bats is having, I would have no conception of what, if anything, I was perceiving; certainly, I would not take myself to be perceiving trees. Presumably the same would go for the bat, if he were suddenly to have experiences like mine. The fact remains that the bats and I do perceive the same trees; as a Kantian might put it, our experiences have a common basis in noumenal reality. It is just that this reality has the power to cause more than one kind of experience: it can produce visual experiences like mine, and echolocational experiences like the bat's.

[2] Actually, Chalmers denies that the existence of the computer (with its powers) is sufficient for the existence of what the brains call "trees," etc. According to him, "[i]f there is an isolated computer simulation to which nobody is connected in [the right sort of] way, we should say that it is simply a simulation." Chalmers (2010a), 470. This strikes me as inconsistent with the spirit, if not the letter, of Chalmers's overall position. What Chalmers should have said, I think, is that an isolated computer is a world of *unperceived* things (including perhaps mountains, lakes, fields, etc.). After all, *our* world would have existed very much as it actually does, even if nobody had existed to perceive any of it—that is, even if nobody had been connected to it in the right sort of way.

Likewise, the computer described in the Vat Hypothesis has the power to produce mountainish experiences in the envatted brains, and computerish experiences in the Vatmaster. In both the bat case and the vat case, the seats of experience-causing power have the power to cause a superabundance of experiences. Within this superabundance we can identify distinct subsets of experiences (envattees' and envatters'; humans' and bats') each of which is best regarded as corresponding to the same reality as that to which the other subset corresponds (computational states also known as mountains; trees also known as whatever the bats might call them). Considered in this light, we have no more reason to deny that the envatted brains perceive trees than to deny that the bats do.[3]

Suppose that Chalmers is right to say that there are trees, lakes, etc. in the vat scenario, despite the fact (as we are assuming it to be) that the vat scenario involves more phenomenological potential than our actual scenario. Would that give us a reason to accept D1 after all?

Maybe. But it would certainly give us a reason to reject D4! Either way, the argument from unusual origins is unsound.

Let me summarize my response to the argument from unusual origins, from the standpoint of radical phenomenalism.

A radical phenomenalist rejects D1, on the grounds that the physical facts in any given situation supervene on facts about *all* the potential experiential voyages that characterize that situation, and not just on a subset of them. Since the potential voyages that characterize our actual situation are only a subset of the voyages that characterize the vat scenario, the fact that there are trees and lakes in the actual scenario does not (according to the radical phenomenalist) imply that there are trees and lakes in the vat scenario.

But suppose that there is a radical phenomenalist who, perhaps inspired by Chalmers, wants to accept D1, holding that the totality of potential voyages that characterizes the vat scenario does bear interpretation as including perceptions of trees. Now, according to a phenomenalist, if it makes sense to speak of things as having an ultimate underlying nature, that nature is of a broadly phenomenological kind. But a phenomenalist might also speak of a thing's *penultimate* nature, consisting of the thing's most fundamental non-phenomenological

[3] It may also be worth pointing out that anyone who denies that the envatted brains perceive mountains, trees, etc. will put himself in the awkward position of committing himself to skepticism about whether he himself perceives such things, since for all we know, the scenario described by the Vat Hypothesis is the scenario that actually prevails! At least, this is true of anyone who is unwilling to accept such aptly-named abominable conjunctions as: "I know it is not a mule, but I don't know it's not a cleverly disguised mule." See DeRose (1995), 27–9, Dretske (2005), Hawthorne (2005), and Collins (2014) for relevant discussion (the mule example is from Collins).

features. Instead of saying that the Vat Hypothesis describes a situation in which there are trees that have an ultimately computational nature, a phenomenalist attracted to Chalmers's view could say that the hypothesis describes a situation in which there are trees that have a penultimately computational nature.[4]

If a radical phenomenalist does say this—that is, if he holds that the totality of potential experiential voyages that characterize the vat scenario is evocative of trees whose existences penultimately reduce to the occurrence of various computational processes—then, like Chalmers, he should reject premise D4 of the argument from unusual origins.

The most natural objection to the argument from unusual origins is that D1 is false. But even if we were to accept D1, it could only be on grounds that compelled us to reject D4. Either way, the argument fails.

9.3 Deceptive appearances

Any metaphysical theory should be able to draw a distinction between veridical experiences and deceptive experiences: between the sort of conscious appearances we normally have while awake and alert, and the sort we have when misperceiving, or dreaming, or hallucinating.

This distinction might seem to pose a special challenge to idealism, since deceptive appearances can, and undoubtedly sometimes do, go undiscovered. But really this is only a challenge to an idealism that attempts to reduce the physical facts to facts about actually existing experience. An idealist theory that proceeds in

[4] In order to say this, the phenomenalist does not have to assume that the vat-totality of potential voyages includes potential voyages of the sort that would characterize the mental life of someone who migrated into or out of the vat, so to speak. No doubt such a migration is possible, in which case there is the potential for a "migratory" experiential voyage in the situation that the Vat Hypothesis describes. We could imagine that there is a Vatmaster who occasionally connects his own brain to the supercomputer, or that he sometimes removes a brain from the vat and inserts it into a body outside the vat. If so, then the migrants will know that trees have a computational underlying nature, but other than that, the situation is much the same as if there were no migration. For even if there weren't, the potential experiential voyages corresponding to the envatted brains and the potential experiential voyages corresponding to the Vatmasters (or possible Vatmasters) would add up to a phenomenological picture of a world in which the underlying nature of trees is computational—or so a phenomenalist might argue.

What if there are trees outside of the computer/vat setup? That seems like a coherent possibility. A Chalmersian phenomenalist could accommodate it by saying that in such a situation, there are two kinds of trees: penultimately computational trees, and trees that are not penultimately computational (unless the vat, the computer, and their surroundings are embedded in a further, second-order envatted-brains scenario). A non-Chalmersian phenomenalist may prefer to say that this is a scenario in which only the trees surrounding the vat are real trees, the "trees" penultimately grounded in operations of the computer being mere pseudo-trees. Whichever story you want to tell, you can tell it in a way that is consistent with radical phenomenalism.

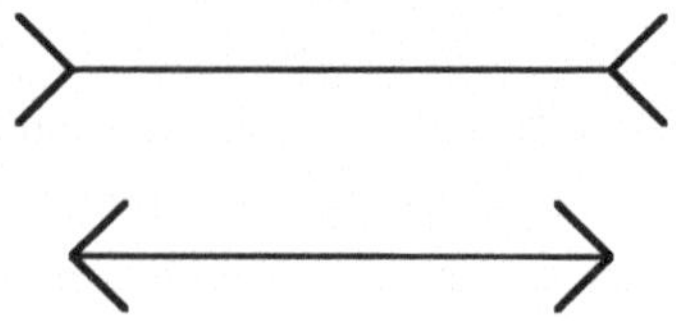

Figure 9.1 Müller-Lyer illusion.

terms of *potential* for experience can account for the distinction between deceptive and veridical appearances straightforwardly, and in a way that allows for the possibility of undiscovered deception.

Consider the experience you have when you look at Figure 9.1. This is an experience as of unequal-length horizontal lines. However, in the phenomenological limit, the experience is not best interpreted as a *perception* of unequal-length horizontal lines. This is because in the limit, there are appearances of someone producing the diagram, appearances of the diagram being measured with a ruler, etc.

Considered on its own, and in isolation from any other possible or actual experiences, the experience you have when you look at the Müller-Lyer is best interpreted as a perception of unequal horizontal lines. But considered in the phenomenological limit, the experience is best interpreted as a perception of equal horizontal lines. This is true even though it is an *appearance* of unequal lines (even in the phenomenological limit), and even if the deception happens to go undiscovered.

Similarly, the experience I have when I mistake a rope for a snake is, sure enough, an appearance of a snake. But in the phenomenological limit, there are also experiences as of a rope being dropped where I now take there to be a snake, as of twisted cotton fibers where I now take there to be snakeflesh, etc. In the light of these experiences, my snakish experience is best interpreted as a misperception of a rope, rather than as a perception of a snake. This is true, even if the corrective experiences are not among the experiences anyone ever actually has; that is, even if my error is never discovered.

Finally, when I dream of a tornado, I have an experience, *E*, as of a tornado. But in the context of all potential experiential voyages, which include experiences as of my brain doing certain things while I'm asleep, as of my brain doing similar things when the eyes to which it is attached are open and directed toward a tornado, etc., *E* is best interpreted not as a perception of a tornado, but as part of a dream about a tornado. If I go my whole life thinking that some event I merely dreamt about actually occurred, and no evidence to the contrary ever comes to light, then the body of actual experience does not demand that we interpret my dream-experience as deceptive. But that doesn't change the fact that the experience is best interpreted as deceptive, in the phenomenological limit.

Thus, I submit that the phenomenalist's best solution to the problem of deceptive appearances is to say that an appearance of a physical object or event x is a deceptive appearance if, and only if, (1) it is an experience as of x, and (2) it is *not* best interpreted as a *perception* of x in the phenomenological limit. This solution not only permits the phenomenalist to maintain a distinction between deceptive and veridical appearances, but also allows him to recognize that some deceptions go entirely undiscovered.

9.4 The imperceptible

We have already seen that phenomenalism can easily account for the existence of unperceived things. But when it comes to things that are not merely unperceived, but imperceptible, the case is not so simple.

To get around the problem of imperceptible things, some idealists go as far as to deny that imperceptible things are possible, or even conceivable. This is not how I want to handle the problem of imperceptibles. I want to grant that there is a distinction to be drawn between physical phenomena that we can perceive and physical phenomena that we cannot, but insist that we can draw this distinction in a way that is both plausible and consistent with phenomenalism.

Kant, unlike Berkeley, holds that there are unperceived physical things. However, there is a sense in which Kant, like Berkeley, denies the possibility of *imperceptible* physical things. For example, although we tend to think that the sub-microscopic parts of physical objects are too small for us to perceive, Kant insists that even these parts are, in a sense, perceptible. For even though we never have *appearances* of, say, an individual sub-microscopic thing,

> the existence of the thing is still connected with our perceptions in a possible experience, and with the guidance of the analogies we can get from our actual perceptions to the thing in the series of possible perceptions. Thus we cognize the existence of a magnetic matter penetrating all bodies from the perception of attracted iron filings, although an immediate perception of this matter is impossible for us given the constitution of our organs. For in accordance with the laws of sensibility and the context of our perceptions we could also happen upon the immediate empirical intuition of it in an experience if our senses, the crudeness of which does not affect the form of possible experience in general, were finer.[5]

Kant's remarks remind us that perceptibility is not a property of a thing, but a relation between a thing and a perceiver (or possible perceiver). A shooting

[5] Kant (1781/1998), 325–6.

star is perceptible to me, but not to a blind man; the Earth's magnetic field is perceptible to a homing pigeon, but not to any man. There is no such thing as simple "perceptibility," unless by this we mean perceptibility by some possible perceiver or other, in which case it is unclear that anything is imperceptible. The most useful notion of perceptibility is perceptibility-to: the shooting star is perceptible *to* me, but not to the blind man, the magnetic field perceptible *to* the pigeon, but not to me.

So it seems Kant recognizes two senses of the word "perceptible."

In one sense, even a magnetic field is perceptible to me: I can perceive it, inasmuch as the field can have some bearing on the character of my mental state (if the field were not present, I wouldn't have precisely *these* appearances of iron filings). In another sense, the magnetic field is imperceptible to me, due to the crudeness of my sense-organs.[6]

My goal in this section is to elucidate the latter sense of "imperceptible," and to give an account of physical phenomena that are imperceptible (to us) in this sense—an account that is both plausible and compatible with a phenomenalist analysis of the physical.

The plausibility requirement is one that historical phenomenalists have struggled to satisfy. In the past, the typical phenomenalist strategy has consisted of two steps: (1) reduce imperceptible physical phenomena to patterns of perceptible physical phenomena; and (2) reduce patterns of perceptible physical phenomena to patterns of conscious experiences.[7] The main problem with this approach is step (1), in which the traditional phenomenalist identifies electrons (for example) with perceptible patterns in the behavior of macroscopic phenomena: galvanometers, glass rods, vacuum tubes, electrical storms, etc. The problem is that such an identification is hard to square with historical scientific developments.

Prior to the invention of the microscope, it was impossible for human beings to perceive microorganisms. Now let's imagine a sixteenth-century phenomenalist who happens to be an early proponent of the germ theory of disease. Let's suppose that this individual takes the standard phenomenalist line on imperceptible things, meaning that he equates the germs that cause plague with the macroscopic symptoms of plague, together with the observable circumstances in which those

[6] Kant himself does not seem to have been very interested in elucidating this second sense of "perceptible" (in which there are many things that are imperceptible to us). According to Langton, Kant tends to work with an idiosyncratic notion of perception, according to which it is enough for me to "perceive" something that the thing has some effect on me, even if it is impossible for me to know that I am so affected; see Langton (1998), 186–9. If we try to interpret Kant as working with a more ordinary concept of perception, we run into the difficulties described in Parsons (1964).

[7] See, for example, Mach (1919), 505–7 and Margenau (1961), 7–8.

symptoms arise. This will prove to be an awkward position for him to sustain when people begin to observe specimens of *Yersinia pestis* under the microscope. A modern-day phenomenalist exposes himself to the danger of a similar embarrassment, if he equates electrons with macroscopic patterns of electrical phenomena.[8]

But there is no need for a phenomenalist to take this approach. Instead of reducing imperceptible things to patterns of perceptible things, and then reducing the latter to something purely experiential, a phenomenalist can reduce imperceptible things directly to patterns of experiences. By this account, perceptible and imperceptible phenomena alike reduce directly to patterns of experiences, where each of the experiences is, perforce, an experience as of something perceptible. The difference between a perceptible phenomenon and an imperceptible phenomenon lies in the nature of the experiential patterns to which they respectively reduce.

More specifically, a perceptible phenomenon reduces to a pattern of experiences that resemble one another in some distinctive phenomenal respect, whereas an imperceptible phenomenon reduces to a pattern of experiences that do not resemble one another in any distinctive phenomenal respect. What defines the pattern of experiences that corresponds to a perceptible thing is the fact that the experiences making up the pattern have some distinctive phenomenal quality in common with one another (or if not a common quality, at least a distinctive family resemblance). What defines the pattern of experiences that corresponds to an imperceptible thing is the fact that its constituent experiences collectively exhibit some more or less abstract regularity (typically expressible in mathematical terms), despite the fact that the experiences do not resemble one another in any distinctive phenomenal respect.

Stated loosely but succinctly, the account of imperceptible physical phenomena that I advance here equates imperceptible entities with *hidden patterns in experience*, as distinct from the manifest patterns in experience that constitute perceptible phenomena.

[8] This criticism of the Renaissance phenomenalist parallels Paul Churchland's criticism of scientific anti-realism, in Churchland (1985), 44–5. It is actually not clear that this kind of criticism constitutes a fatal objection to the usual phenomenalist line. A phenomenalist of the traditional stripe could argue that the advent of the microscope simply expanded the range of macroscopic plague symptoms to include sightings of *Yersinia pestis* under the microscope. Alternatively, he might argue that, unlike individual specimens of *Yersinia pestis*, individual electrons are *in principle* immune to direct observation—or at least, that individual electron-states have this kind of immunity—and therefore could not cause the kind of embarrassment that the Renaissance phenomenalist felt vis à vis microscopic observations of bacteria. For better or worse, opponents of the standard phenomenalist approach appear to be unmoved by such arguments.

In order to make the account plausible, we need to ask why we draw the distinction between perceptible and imperceptible (or observable and unobservable) phenomena in the first place. Only after we have answered this question will we be in a position to judge whether the account I offer does justice to the distinction.

Since it goes without saying that there are perceptible things, asking why we distinguish between perceptible and imperceptible things is tantamount to asking why we think that there are imperceptible things. Why do we believe that there is more to the physical world than meets the eye?

The reason why we believe this is that we are able to *explain* perceptible phenomena in terms of imperceptible phenomena. Indeed, to a decent first approximation, natural science just is the practice of explaining the perceptible in terms of the imperceptible. Were it not for our attachment to this practice, we would have no use for imperceptible things.

So what is it, to explain something perceptible in terms of something imperceptible? Here we can do no better than to look at some of the examples of such explanations that history affords.

One example we've already mentioned: explaining the spread of disease by reference to the transfer of microbes. Two classic cases in point are Ignaz Semmelweis's study of puerperal fever at the Vienna General Hospital in 1847, and John Snow's analysis of the London cholera outbreak of 1854.

Semmelweis noticed that the incidence of puerperal fever among women who gave birth in one of the two clinics he managed was far higher than in the other clinic. This was puzzling, since the clinics were located in the same geographical area and employed the same obstetric techniques. Through a process of elimination, Semmelweis came to focus on the fact that the high-incidence clinic employed medical students, while the low-incidence clinic employed only nurses and apprentice midwives. The breakthrough came when Semmelweis turned his attention to what these two groups of medical workers were doing beyond the obstetric ward. Crucially, the medical students, but not the midwives, performed autopsies as part of their training. Semmelweis hypothesized that the medical students were transferring an imperceptible cadaveric pathogen from the bodies they dissected to the patients in the obstetric ward, via unwashed hands. This hypothesis was confirmed when he required the students to wash their hands before entering the ward, with a consequent 90 percent drop in the incidence of puerperal fever in the problem clinic.[9]

John Snow confronted a similar challenge in 1854, while trying to pin down the source of a severe outbreak of cholera in the Soho district of London. At

[9] See Pelczar et al. (1993), 591–4.

the time, the received view was that cholera outbreaks were due to the spread of "bad air," identifiable by its foul odor. Skeptical of this theory, Snow made a map of households affected in the outbreak, and found that almost all of them were closer to a particular water pump (the infamous Broad Street pump) than to any other water source. The exceptional households proved to be ones with children who attended school near the Broad Street pump, or with adults who used the Broad Street pump in preference to nearer sources of water. Snow hypothesized that the cholera outbreak was due to the presence of an imperceptible pathogen in the water dispensed by the Broad Street pump. This hypothesis was confirmed by the revelation that the employees of a brewery located near the pump were untouched by the outbreak: they drank only beer made with water from the brewery's private well.[10]

Today, of course, it is possible to perceive the infectious agents posited by Semmelweis and Snow using microscopes more powerful than those available to these pioneers of epidemiology. But taken in historical context, these are classic examples of explanations of the perceived in terms of the imperceptible. What I would like to point out here is that the explanations have a similar structure.

In each case, what calls for explanation is a perceptible phenomenon, an observed pattern of people getting sick. Also in each case, the sick people do not differ in any relevant perceptible respect from many other people who are not getting sick: the bad clinic and its inmates are in all relevant respects observationally indistinguishable from the good clinic and its; the people and provisions in the cholera-stricken households are in all relevant respects observationally indistinguishable from those in nearby households free from cholera. The cases of illness have no distinctive perceptible feature in common (besides the symptoms of the illness itself) to distinguish them from cases in which illness is absent.

Nevertheless, taken collectively, the cases of illness do instantiate a certain pattern. The hands that deliver babies in the bad clinic *look* the same as the hands that deliver babies in the good clinic, but only the former hands have performed autopsies earlier in the day. The water in the cholera-stricken household's jug *tastes* the same as the water in the unstricken household's jug, but only the former water came from the Broad Street pump. In both cases, there is a pattern to the illness, but it is not a pattern that meets the eye.

Another classic example of explaining the perceptible in terms of the imperceptible comes from the field of chemistry. It is possible to classify chemical elements into groups, where each group contains elements that resemble one another in terms of their chemical behavior (how readily they interact with various other

[10] For details on the Broad Street incident, see Vinten-Johansen et al. (2003), 283–317.

elements to form compounds). However, the elements that belong to a given group do not have any distinctive perceptible qualities in common that serve to explain the characteristic chemical behavior of the elements in that group. In some cases, the elements in a given group bear little outward resemblance to one another at all. For example, oxygen, sulfur, and polonium all belong to the same chemical group, despite the fact that at normal temperature and pressure, oxygen is a colorless gas, sulfur a brittle, yellow, non-metallic solid, and polonium a silvery radioactive metal.

If anything explains why the elements in a given group exhibit similar chemical behavior, it is not, therefore, that those elements have some perceptible quality in common—some observable property that distinguishes them from elements in other groups. Rather, if there is an explanation, it must be that the elements in a group all share some distinctive *imperceptible* quality.

The first person to put his finger on this quality was Dmitri Mendeleev. Mendeleev found that if you tabulate the elements in order of their experimentally-established atomic weights, starting a new table row when the elements' experimentally-established valences begin to repeat, each column of the resulting table contains all the elements of a single chemical group. (The valence of an element is the number of atoms that a single atom of that element can bond with at one time.) This led to the hypothesis that the chemical behavior of an element depends on some imperceptible feature of the atoms of that element, where this feature is a function of the element's atomic weight and valence. This imperceptible feature is now known to be the number of electrons that an atom of an element contains in its outermost electron shell.[11]

Mendeleev's explanation of the chemical behavior of different elementary groups has the same general structure as that of Semmelweis's explanation of puerperal fever, and Snow's explanation of the London cholera outbreak. In each case, the explanation consists of identifying a pattern in the physical phenomenon to be explained that is not reflected by any distinctive *apparent* similarity among the different instances of that phenomenon. In the biological cases, this was a covert demographic pattern; in the chemical case, it is a covert numerical pattern.

We find the same structure of explanation in physics. When Newton explained the motions of the Moon, the planets, and falling apples in terms of a propensity for massive bodies to accelerate toward one another in proportion to their masses, and in inverse proportion to the square of the distance separating them, he was not recording a humanly perceptible pattern of events. The Moon, planets, and

[11] The story behind the periodic table of elements is actually considerably more nuanced than these comments suggest; see, for example, Scerri (2009).

falling apple do not *look* like they are conforming to an inverse square law of attraction. If they did, there would be no need for physics: we would all be natural-born Newtons.

The pattern that Newton discovered does not consist of the repeated instantiation of some distinctive perceptible quality that lunar, planetary, and terrestrial motions all have in common. Rather, it is a pattern to which these objects conform despite *not* resembling one another in any distinctive perceptible respect. It is a mathematical pattern, which Newton was able to discern only by analyzing the terrestrial and astronomical observations of his predecessors and contemporaries.[12]

How should we describe Newton's explanation of gravitational phenomena, from the standpoint of radical phenomenalism? Massive bodies' conformity to an inverse square law of attraction is imperceptible, in the sense that (1) our appearances of bodies are (in the limit) interpretable as perceptions of bodies conforming to such a law, *despite* the fact that (2) these appearances do not resemble one another in any distinctive phenomenal respect. (2) is just a fact about actual human experience. We have experiences *of* bodies conforming to the law of gravitation, but we do not have experiences *as of* bodies conforming to that law. If we did, our perceptions of the apple, planets, and Moon would have a phenomenal similarity that they do not, in fact, exhibit.

Newton's achievement was to establish (1), by actually providing the mentioned interpretation. Of course, in all likelihood, Newton did not take himself to be interpreting a collection of conscious experiences. He probably took himself to be interpreting a collection of physical states of affairs, thought of as the causes of those experiences. But even if this is how he thought about it, it is not how *we* have to think about it. We can accept Newton's mathematical interpretation of the relevant astronomical observations, without accepting his metaphysical interpretation of them.

In effect, Newton made two discoveries at once. He discovered that physical objects are subject to an inverse square law of attraction. And he discovered that conscious experiences are interpretable as including perceptions of physical objects that are subject to an inverse square law of attraction. This much is uncontroversial. The question that confronts the metaphysician is: What is the logical relationship between these two discoveries?[13]

The man in the street will say that the fact that the objects obey the law *explains* the fact that our experiences are interpretable as including perceptions of objects

[12] Chief among them Galileo Galilei, Tycho Brahe, and John Flamsteed (first Astronomer Royal).

[13] The latter discovery was a discovery of what I have called a law of experience (see Chapter 5, Section 3)—or rather, it *would* have been the discovery of a law of experience, had the inverse square law proved to be a true law of nature.

obeying the law. A Kantian will say that the fact that the objects obey the law *reduces to* the fact that noumena tend to cause experiences that we can interpret in the way described. A phenomenalist says that the fact that the objects obey the law reduces to the fact that the experiences are (in the phenomenological limit) interpretable in the described way. The point of the present discussion is not to settle this debate, but only to make it clear that it is possible to understand Newton's achievement—his explanation of the observable motions of bodies in terms of an invisible force—in a way that is consistent with phenomenalist principles.

I suggest that what is true of the Newtonian case is true of all cases in which we successfully explain an observable phenomenon in terms of something unobservable. Scientists explain the behavior of lightning, galvanometers, and tumble-dried laundry in terms of the behavior of electrons; this is an explanation of the perceptible in terms of the imperceptible. What does the explanation come to? I suggest that it comes to the fact that we can interpret our conscious appearances of lightning, galvanometers, and laundry as perceptions of things that consist of particles which behave in ways that satisfy Maxwell's equations of electrodynamics, even though these conscious appearances have no qualia in common that distinguish them from all other appearances. There is, for us, no such thing as an experience as of physical phenomena satisfying a set of partial differential equations. But that does not imply that our appearances of lightning and the rest are not interpretable (in the limit) as experiences *of* phenomena satisfying these equations. They are so interpretable, as Maxwell demonstrated.

The fact that the existence of an electron reduces to the existence (in the limit) of a hidden pattern in experience does not make the electron any less real than a billiard ball. For the existence of the billiard ball too reduces to the existence (in the limit) of a pattern of experiences—it is just that this pattern, unlike the pattern that constitutes an electron, is defined by phenomenal resemblances among the experiences that constitute it.

In the same way, reducing bodies' propensity to move in accordance with the law of gravity to a hidden pattern in our appearances of physical objects does not make gravity less real than apples, planets, or the Moon. For the Moon too exists as a pattern in the phenomenological limit: it is just that this pattern, unlike the gravitational pattern, is defined by the fact that its constituent experiences bear a distinctive phenomenal resemblance to one another: they are all appearances of the Moon.

In Sellarsian terms, imperceptible phenomena correspond to regularities that belong to conscious experience, without, however, belonging to the "manifest image" of the world. Unobservable things are, as it were, steganographic messages, woven into the pattern of experience like a text in Morse code woven into the pattern of a scarf.[14]

The phenomenalist account of imperceptible things has a special advantage, in that it sheds light on the different ways in which technological advances facilitate scientific progress.

Intuitively, the advent of the microscope had a different kind of impact on scientific inquiry from the development of the astrolabe. A microscope facilitates a certain phenomenal unification of our perceptions of various macroscopic phenomena. Before the microscope, human perceptions of cholera had nothing phenomenal in common, beyond all being appearances of a characteristic set of symptoms. Since the microscope, human perceptions of cholera *have* shared a common phenomenal quality: they have been appearances of a certain bacterium viewed under a microscope. (At least, this is true of some humans' experiences.) It is not too much to say that the microscope has expanded our perceptual horizons.[15]

The impact of the astrolabe was quite different. This naked-eye device, used to measure the position of a celestial body in the sky relative to a point on the horizon, does not allow us to perceive anything that we can't perceive without it. What it allows us to do is to record our perceptions in a systematic way over extended periods of time, for subsequent tabulation and mathematical analysis. If John Snow had viewed water samples from the Broad Street pump under a modern microscope, he would have perceived that the cause of the cholera outbreak was a tiny living thing. By contrast, no amount of stargazing—with or without an astrolabe—can result in our perceiving that the planets sweep out equal areas in equal times as they pursue elliptical orbits around the center of gravity of the Solar System. The microscope unifies experience in a way that is open to view and easy to discern; the astrolabe unifies experience in a way that only arduous calculation can establish. The astrolabe expanded our cognitive horizons, but not by expanding our perceptual horizons.

Let's return to the phenomenalist analysis of the physical, and see what we must do to make it compatible with the existence of imperceptible physical phenomena. As we left it in Chapter 8, the analysis was not compatible with this, stating (roughly) that a physical thing exists if, and only if, the totality of potential experi-

[14] Seen this way, the manifest and scientific images are not rivals; if there is a rivalry, it is between those who claim to find a hidden message in the scarf and those who refuse to see in it anything more than a pretty pattern.

[15] Whether this expansion should be understood literally is a topic of debate in the philosophy of science; see, for example, Hacking (1985), Churchland (1985), and Fraassen (1985).

ences is, in the phenomenological limit, interpretable as including a *perception* of that thing. To get an analysis that allows for the existence of imperceptible physical things, we must amend the earlier analysis to read as follows:

For all x, x exists in spacetime if, and only if, (1) the probability that S_n evokes x approaches 1, as n approaches infinity, or (2) the probability that S_n contains a hidden pattern of experiences answering to x approaches 1, as n approaches infinity,

where, as before, "S_n" stands for a random set of n potential experiential voyages. This is just the conjunction of the local analysis from Chapter 8 with the phenomenalist account of imperceptible things given above.

9.5 Conclusion

Over the course of the last three chapters, I have developed a comprehensive phenomenalist theory, and defended that theory against a variety of objections. To overcome these objections, a phenomenalist must: (1) uphold the distinction between deceptive and veridical experiences; (2) account for the reality of unperceived and imperceptible things; (3) allow that physical things do not, in general, depend for their reality on the existence of minds or experiences; (4) make sense of the order that characterizes our conscious mental lives; and (5) carry out a phenomenological reduction of physical time (in addition to physical space).[16]

In this chapter, I have argued that a phenomenalist can achieve (1) and (2). In Chapter 7, I argued that a phenomenalist can achieve (3) and (4). In Chapter 8, I argued that he can achieve (5).

Radical phenomenalism does not conflate physical reality with a dream from which we never wake up. It does not rule out the possibility or reality of unperceived or imperceptible things. It does not entail a mind-dependent conception of the physical world. It does not make the orderliness of conscious experience any more mysterious than any other theory (or if it does, it is only in exchange for making other things correspondingly less mysterious). It does not leave facts about time and the temporal unreduced.

Even the most obdurate opponents of phenomenalism will concede that phenomenalism is nice work, if you can get it. The complaint against phenomenalism was never that it promised too little, but that it could not make good on the much that it does promise. Speaking of Berkeley, Peirce comments that "[t]he few who do now care for metaphysics are not of that bold order of minds who delight to hold a position so unsheltered by the prejudices of common sense as

[16] See Armstrong (1961), 47–79.

that of the good bishop." But what Peirce calls prejudices are, of course, really just objections, and those who hold a position in the face of unanswered objections are not so much bold as rash. My hope in answering the principal objections to radical phenomenalism is that this will allow at least some minds to take some delight in holding a position that is, if anything, even bolder than the good bishop's.[17]

[17] For Peirce's remarks, see Peirce (2009), 75.

10

The conscious self

> When philosophers use a word—"knowledge," "being," "object," "I," "proposition," "name"—and try to grasp the *essence* of the thing, one must always ask oneself: is the word ever actually used this way in the language-game which is its original home?
>
> Ludwig Wittgenstein, *Philosophical Investigations*

10.1 Introduction

The preceding three chapters have developed and defended the central position of this book, a radical phenomenalism that analyzes facts about spacetime and its contents into facts about conscious experience. The goal of this chapter and the next is to deepen the phenomenalist position by arguing that a phenomenalist need not regard either the conscious self (subject, mind) or mental representation (intentionality, content) as an explanatory primitive.

It is possible to adopt a version of phenomenalism that is primitivist in either, or both, of these respects. Following Leibniz, we could take conscious selves as primitive and irreducible objects; following Brentano, we could take intentionality as a primitive and irreducible property or relation. But other things being equal, it is preferable to have a theory in which minds and mental representation are derivative phenomena, rather than components of the ontological bedrock.[1]

The goal of this chapter is to show that there is no need for a radical phenomenalist to take conscious selves as conceptual or explanatory primitives.[2]

If every conscious mental life consisted of a single stream of consciousness, we would already have reached this goal, since we have already given (in Chapter 5) an

[1] For Leibniz's primitivism about minds, see Leibniz (1714/1989a), Section 3. For Brentano's primitivism about mental representation, see Brentano (1874/1995), 68.

[2] Throughout this chapter, I use the phrases "conscious mind," "conscious self," and "conscious subject" interchangeably. I also use the phrase "conscious mental life" in a way that makes it a definitional truth that two experiences occur in the same conscious mind (self, subject) if, and only if, they belong to the same conscious mental life.

analysis of the stream of consciousness that is compatible with radical phenomenalism. But just as the experiences that make up a stream of consciousness do not do so by belonging to a single experience, the streams of consciousness that make up a conscious mental life do not do so by belonging to a single stream. So an account of the conscious mind does not come for free with an account of the stream of consciousness. The question remains: how can experiences that do not belong to the same stream of consciousness belong to the same subject of experience?

This is different from asking: how can experiences not belonging to the same stream of consciousness belong to the same *person*? The set of all people is distinct from the set of all selves. A squirrel is a conscious self—it has a mind, it is a subject of experience—but it is not a person. Whether there could be a person who was not a conscious self is less clear; maybe a zombie would qualify (certainly, a zombie could have a *personality*). It does not seem implausible to think that a single conscious self might take the form of distinct people at different phases of its existence, and it might even be possible for a single person to host distinct conscious selves at different stages of his life.

Our concern here is selfhood, not personhood. Specifically, we are interested in *conscious* selfhood. Maybe there are other forms of selfhood. For example, maybe there is something we could call "psychological selfhood." Maybe a zombie is a psychological self (with beliefs, desires, etc.), despite not being a conscious self. Our question here is: how can experiences that do not belong to the same stream of consciousness belong to the same *conscious* self?[3]

There are four ways you could respond to this question.

First, you could respond by saying that *belonging-to-the-same-conscious-mental-life-as*—"co-subjectivity," for short—is just a primitive relation, a conceptual and explanatory fundamental that we cannot usefully define in terms of anything more basic. This is the Leibnizian *primitivism* we are hoping to avoid.

Second, you could respond by denying that it is possible for experiences in different streams to belong to the same mental life. This is *anti-realism* about the (multi-stream) conscious self.

Third, you could respond by offering a reductive analysis of co-subjectivity, specifying criteria for the obtainment of this relation between any two experiences (including experiences in different streams). This is *reductionism* about the conscious self.

Fourth, you could respond by holding that there is no such thing as *the* relationship by virtue of which two experiences belong to the same conscious

[3] Hereafter, I use "self," "subject," and "mind" as short for "conscious self/subject/mind."

mental life, but different relationships that we refer to on different occasions when speaking about conscious subjects, minds, selves, or lives. This is *particularism* about the self.

In the next section, I briefly discuss anti-realism about selves that span multiple streams of consciousness. In Section 10.3, I review and criticize the main reductionist analyses of the conscious self. Finally, in Section 10.4, I put forward a particularist view of the self.

10.2 No self?

Some people deny that there are any conscious selves at all. Their position is, to put it mildly, implausible. But one could deny the reality of selves that include multiple, complete streams of consciousness without denying the reality of *all* conscious selves. There are two reasons one might take this view.

First, you might hold that what we ordinarily think of as interruptions in the stream of consciousness are no such thing. This appears to have been Leibniz's view, although it has few adherents today. On this view, so-called dreamless sleep is not a period of unconsciousness separating distinct streams of consciousness. Rather, it is an unapprehended and unremembered portion of a very long stream of consciousness that extends from birth to death (or, on Leibniz's view, from eternity to eternity).[4]

Alternatively, you might concede to common sense that the onset of sleep usually does bring an end to a stream of consciousness, but insist that the end of the stream marks the end of a conscious mental life—the end of a conscious self. On this view, what we normally think of as multiple streams of consciousness within a single mental life are really distinct mental lives.[5]

Neither of these positions is very attractive. But there is no need for us to consider them further here. This is because if there are no multi-stream selves—that is, if experiences belonging to different streams of consciousness never belong to the same conscious mental life—then a radical phenomenalist has no obligation to account for such selves. As noted earlier, a phenomenalist already has a way to account for streams of consciousness; if, as the anti-realists maintain, the only sort of self that exists just *is* a stream of consciousness, then the phenomenalist account of the stream of consciousness is de facto an account of the self.

[4] See Leibniz (1714/1989a), 214–16. For a recent defense of a broadly Leibnizian position, see Gustafsson (2011).

[5] See Strawson (1999), 100–1.

10.3 Reductive theories of the conscious self

The most common stance on the conscious self, at least in the modern Western philosophical tradition, is reductionism. Reductive theories fall into three categories:

(1) *substrate theories*, according to which experiences in different streams belong to the same mental life to the extent that the experiences are caused or realized by a single object that we can understand in purely non-mental terms (e.g. by a single functionally- or physiologically-individuated brain);
(2) *asymmetric dependence theories*, according to which experiences in different streams belong to the same mental life to the extent that one of the streams depends on the other for one or more of its phenomenological features (but not vice versa); and
(3) *hypothetical-stream theories*, according to which experiences in different streams belong to the same mental life to the extent that the experiences' respective causes or realizers would have caused or realized experiences belonging to a single stream of consciousness, had all potential for conscious experience been actualized.

The debates surrounding these theories have not taken place against the background of a phenomenalist metaphysics, and there would be little point in restating them in a phenomenalistically acceptable way. My goal in this section is to show that the theories all fail, and it will be enough for this purpose if I can show that they fail even on the anti-phenomenalist assumptions of their proponents. So in this section, we will relax our phenomenalist principles, and conduct the discussion in more conventional terms.

One thing that will emerge from our discussion of the analyses listed above is that every one of them entails a graded conception of the self, according to which the inclusion of two experiences in the same conscious mental life is not always a simple yes-or-no matter, but a matter of degree. That is why the phrase "to the extent that" occurs in the analyses, instead of "if and only if."

I propose to assess substrate, asymmetric dependence, and hypothetical-stream theories by reference to four test cases. Each of these is a case in which two experiences intuitively do, or, as the case may be, intuitively do not, belong to the same conscious mind (the same conscious mental life). A successful account of the conscious self should confirm our intuitions about these cases. The cases are:

Jumping Streams
We are standing some distance from one another, admiring an apple tree, when suddenly I experience a "jump" from my original perspective to your original perspective, and vice

versa: I now see the tree from your old angle, and you see it from my old angle. Neither of us loses consciousness: each of us has an experience as of viewing the tree from one angle and then another (and not just an experience as of viewing the tree from one angle, and then an experience as of viewing the tree from another angle). I continue to have conscious experience in this vein, including sensory-somatic experiences of what used to be your body, and vice versa for you. In short: the organism that used to embody my stream of consciousness now embodies your stream of consciousness, and the organism that used to embody your stream of consciousness now embodies mine.[6]

Parallel Production

For each of my conscious experiences, *E*, there is another conscious experience phenomenally indistinguishable from *E* that has the same neural correlate as *E*, but occurs in a different conscious subject from *E*, as part of a distinct stream of consciousness. If we want to think of this in causal terms, we can say that every brain-event that causes an experience in my stream of consciousness also causes an exactly similar experience in a parallel stream of consciousness. In short, my brain sustains two concurrent conscious mental lives, rather than just one.

Two Brains

There are two envatted brains, Alpha and Beta. Both brains are connected to a computer, which is programmed to stimulate the brains as follows. The computer begins by stimulating Alpha in such a way that Alpha has an ordinary stream of consciousness, much like the one that you are in the midst of right now. At the same time, the computer monitors Alpha and records a detailed description of Alpha's experiences on a separate hard disk. The next day, the computer gives Beta, which has never had experiences anything like Alpha's, a stream of consciousness that includes conscious memories, or apparent memories, of experiences just like the ones that Alpha had the preceding day. (The computer knows which apparent memories to induce in Beta by consulting its stored description of Alpha's experiences from the preceding day.) In short, the computer is so programmed that it gives Beta an apparent memory of having had an experience *E* on the preceding day only on condition that *Alpha* did in fact have an experience just like *E* on the preceding day. Despite this peculiar arrangement, the brains give rise to experiences in distinct conscious subjects.

Hostage Situation

You have been taken hostage by a wealthy madman who threatens to kill you if you do not fall asleep before sunrise, but promises to set you free with a million dollars if you do fall asleep before then. There is no doubt that he will follow through with this threat (or promise). Fortunately, you do manage to fall asleep, despite the high anxiety of the situation. When you wake up, your captor is gone, the door to the motel room where you were being held stands open, and there is a briefcase full of hundred-dollar bills on the table next to you. You are glad to be alive, and wonder whether the police will let you keep the money when you report the incident to them.

[6] Compare with the discussion of "streamal diverters" in Dainton and Bayne (2005), 556.

We now consider substrate, asymmetric dependence, and hypothetical-stream accounts in the light of these cases. As we shall see, no account does justice to every case.

10.3.1 Substrata

The substrate theory's analysis of the conscious self is probably the most commonsensical of the three accounts we are considering. The account fits well with the idea that experiences that belong to the same conscious mental life are just experiences that arise from the same brain. That said, the substrate theorist is not committed to equating conscious selves with brains. What makes two experiences belong to the same conscious self, according to the substrate theory, is that the experiences arise from or get realized by the same entity. Whether the entity is a brain, a computer, or something else is unimportant.[7]

The substrate theory implies that co-subjectivity is a matter of degree. This is because it is a matter of degree whether two functional states are states of the same functional system, and a matter of degree whether two neural states are states of the same brain. If I install a bill acceptor on a soda machine that previously accepted only coins, it seems most natural to say that I have upgraded the system, not replaced it. The system survives the upgrade. But if I laboriously convert the soda machine into a dishwasher, it seems most natural to say that the dishwasher replaces the soda machine; the original system does not survive the conversion. In between these extremes is a continuum of cases. Toward the middle might be a case in which I convert the soda machine into a jukebox, or into a vending machine that dispenses live fish bait.

Similarly, a human skull typically contains a single physiologically-individuated brain throughout its existence, even though that brain undergoes many physiological changes. But we can imagine a spectrum of cases in which it is decreasingly natural to say that a brain has survived some changes. For example, we can imagine that scientists map your brain in detail, then break it down into its constituent cells, which they keep alive somehow. If they later reassemble the cells exactly according to their map, it is very natural to say that the brain they put together is the same one they took apart. But if they reassemble the cells according to a detailed map they made of a chimpanzee's brain, it is very natural to say that the brain they put together is *not* the same one they took apart. In the middle of the spectrum, the scientists might assemble the cells according to a heavily modified but still human version of the detailed plan that they made of your brain prior to disassembly.

[7] Substrate theories are also known as "somatic" theories, since the chosen substrate is almost always a physiologically-individuated organism. For recent defenses of the substrate theory, see Olson (1997) and Mackie (1999).

The substrate theory tests well against *Two Brains* and *Hostage Situation*. The theory correctly says that the experiences associated with the different brains belong to different conscious subjects. It also says, correctly, that you survive the hostage ordeal.

But the substrate theory fails the *Jumping Streams* and *Parallel Production* tests. The experiences I have after the jump are assuredly mine: they belong to the same conscious self, and indeed to the same stream of consciousness, as the experiences I had before the jump. Yet, the experiences I have before and after the jump arise from different brains, or get realized by different functional systems. The substrate theory therefore implies, wrongly, that my post-jump experiences do not belong to the same mental life as my pre-jump experiences. The theory also wrongly implies that all of the experiences in the *Parallel Production* scenario belong to the same conscious mental life, since they arise from the same brain, and correspond to states of the same functional system.

10.3.2 Asymmetric dependence

The most popular analyses of the conscious self rely on the idea that one mental state can depend on another, prior mental state in a way that justifies counting both mental states as states of the same mind. This is what I have called the asymmetric dependence theory.[8]

In its classic form, the theory identifies episodic memory as the asymmetric dependence relation that holds disparate mental states together into a single mental life.

Memory is primarily a psychological concept, and traditional memory theories are, first and foremost, theories of what I have called the "psychological self"—theories that apply not just to conscious beings like us, but to unconscious beings that exhibit a sufficiently impressive array of cognitive capacities (such as zombies). But that need not prevent a traditional memory theorist from giving an answer to the question of how experiences in different streams of consciousness can belong to the same conscious mental life. A memory theorist can answer that experiences e_1 and e_2 (occurring in different streams) occur in the same conscious subject if, and only if, (1) the stream to which e_2 belongs contains a memory of an experience in the stream to which e_1 belongs, or (2) the stream to which e_2 belongs contains a memory of an experience in a stream that contains a memory of an experience in the stream to which e_1 belongs, or (3) the stream to which e_2 belongs contains a memory of an experience in a stream that contains a memory

[8] Perhaps the most influential development of the asymmetric dependence approach is Parfit (1984).

of an experience in a stream that contains a memory of an experience in the stream to which e_1 belongs, etc.[9]

While memory has been the relation of choice for traditional asymmetric dependence theorists, it is not clear that it is necessary, or even advisable, for an asymmetric dependence theory of the conscious self to restrict itself to memory.

Our past experiences inform the quality of our present experiences in all kinds of ways. Had I never heard this tune before, it would not sound the same to me now. Had I never driven down this road before, it would not look the same to me now. Had I never tasted coffee before, this mouthful of it would not taste the same to me now. No doubt memory is a prominent manifestation of the dependence of one conscious mental state upon another, but it is not the only manifestation, and not strictly necessary to hold distinct streams together into a conscious self. It is enough if the phenomenology of one stream informs the phenomenology of another stream in some non-memorial way of which the conscious subject may not even be aware (such as one of the ways described above). Indeed, non-memorial dependence relations between experiences might be all that holds together the conscious mental lives of certain lower animals.[10]

How does the asymmetric dependence theory fare on our four tests? It passes *Jumping Streams*, *Parallel Production*, and *Hostage Situation* without difficulty. My experiences after the jump do include conscious memories of experiences from before the jump. Neither of the streams of consciousness in *Parallel Production* asymmetrically depends for its phenomenology on the phenomenology of the other stream. When you wake up after the hostage ordeal, you have vivid memories of the events of the preceding day.

But the asymmetric dependence theory falters on *Two Brains*. In this scenario, Beta has conscious apparent memories of having had experiences that it never really had, but that Alpha did have. Moreover, the fact that Beta has the memorial experiences that it has today heavily depends on the fact that Alpha had the perceptual experiences it had yesterday. This dependence is, furthermore, asymmetrical: the phenomenology of Beta's present stream of consciousness depends on the

[9] Recursively: if, and only if, e_1 belongs the set of experiences S that contains just (1) e_2, (2) any experience that is represented in memory by a member of S, and (3) any experience that belongs to the same stream of consciousness as a member of S. This basic account would have to be elaborated in various ways, to handle the possibility of surviving various kinds of memory loss, brain-splitting, etc.; the available moves are well known, and there is no need to rehearse them here. See, for example, Perry (1976), Parfit (1984), Shoemaker (1984), and Campbell (2001).

[10] O'Shaughnessy (2000), 55–7 suggests that any experiential dependence that can unify the disparate experiences of a mental life should be considered a form of memory. I have no objection to calling the dependence a kind of memory, as long as we don't let the word "memory" saddle the asymmetric dependence account with extraneous psychological baggage. For more on the question of what we should count as "memory" in this context, see Phillips (2010), 193–5.

phenomenology of Alpha's past stream, but not vice versa. The asymmetric dependence theory therefore implies that the conscious experiences presently associated with Beta are co-subjective (to a very high degree) with the conscious experiences that arose yesterday in connection with Alpha. This is the wrong result.

The *Two Brains* scenario is simply a distillation of innumerable counter-examples to the asymmetric dependence theory that it is possible to cook up. Devising such a counter-example is merely a matter of imagining a suitably deviant causal or quasi-causal pathway leading from the phenomenal contents of one mind to the phenomenal contents of another. Since such pathways can cut across distinct conscious mental lives in any number of ways, the asymmetric dependence theory fails as an attempt to cut phenomenological reality at the joints.

10.3.3 Hypothetical streams

Consider a series of streams of consciousness existing between t_0 and t_n, t_0 being the time of occurrence of the first experience in the first stream in the series, t_n the time of occurrence of the last experience in the last stream. According to the hypothetical-streams analysis of the conscious self, these streams belong to a single conscious life if, and only if, the producers of their constituent experiences would have produced experiences belonging to a *single* (long) stream of consciousness, had every experience-producer existing between t_0 and t_n actually produced experiences.[11]

Why *every* experience-producer existing between t_0 and t_n? Why not say that two experiences belong to the same self as long as the brain that causes them would have caused co-streamal experiences, had that brain actually produced experiences continuously from t_0 to t_n? Because then the hypothetical-stream theorist would owe us an account of which brain "that" brain is, without falling into circularity by identifying it as the brain that causes the experiences of a given *self*. But what other remotely relevant way is there to individuate brains, other than functionally or physiologically? Apparently none. That is why we must formulate the hypothetical-stream theory in terms of "*every* experience-producer": otherwise, the theory simply becomes a substrate theory.[12]

[11] Of the three reductive theories of the conscious mind we are considering, the hypothetical-stream theory is the newest, developed primarily in Dainton (2008), 75–114 and Dainton and Bayne (2005), 564–8. Mill suggests an account of the conscious self similar to Dainton's: see Mill (1865/1979), 190.

[12] For the same reason, a hypothetical-stream theorist cannot say that two experiences belong to the same self just in case they arise from the same disposition to generate conscious experiences; the question would then become: under what circumstances does one experience arise from the same generative disposition as another? Substrate theories that emphasize the importance of conscious experience to the self are what Bayne calls "substrate phenomenalism"; see Bayne (2010), 287. The term "co-streamal" is Dainton's.

The hypothetical-stream theory tests well against *Jumping Streams, Parallel Production,* and *Two Brains*. The experiences in *Jumping Streams* do actually belong to the same stream of consciousness; the pairwise produced experiences in *Parallel Production* belong to different streams (and different subjects) even if all potential for experience is being realized; no matter how thoroughly the computer might stimulate the brains in *Two Brains,* Alpha's experiences would occur in a different stream of consciousness from Beta's.

But the hypothetical-stream theory fails against *Hostage Situation*. In this scenario, it is false that if all potential causes of experience had produced experiences all night long, then whatever produces your waking experiences would have produced experiences co-streamal with experiences from the previous day. For in this scenario, if you had not lost consciousness, then whatever actually produces the experiences you have when you wake up would not then have produced any experiences at all (since the madman would have killed you). But according to the hypothetical-stream theory, that means that the experiences you have when you wake up do not occur in the same mind as the experiences you had before you fell asleep.[13]

Nor is this a terribly unrealistic kind of case. It might have been that I *had* to sleep last night in order to survive until today. This might have been the case if I hadn't slept for a long time before, or if doctors had to put me into a medically induced coma in order to save my life.[14]

[13] We can also imagine that if I had stayed up all night, my stream would have jumped to your brain, while your stream jumped to mine. In that case, what does the hypothetical-stream theory say about my present experiences—the ones that I'm having now after a good night's sleep? Would the brain that is producing these experiences be producing experiences co-streamal with those that I remember having yesterday if all brains had produced experiences all night long? No: it would be producing experiences co-streamal with experiences that only *you* can remember from yesterday. The hypothetical-stream theory therefore requires us to conclude, incorrectly, that in this scenario, my present conscious experiences belong to a mental life completely different from the life to which they do, intuitively, belong.

[14] In Dainton (2008), 11–14, Dainton introduces a thought-experiment intended to support a phenomenal conception of the self over a psychological conception. In this thought-experiment, you are anaesthetized and your body deep-frozen, then sliced into thin cross sections that are scanned to yield a detailed cell-level map of your body. The cross sections are destroyed, and the map later used to construct myriad copies of you in a future where such a procedure is straightforward. These copies are psychologically continuous with you (the person who was sliced up), but intuitively none of them *is* you. The phenomenal account respects this intuition, since none of the copies would have had experiences co-streamal with experiences produced by your pre-sliced brain, had all experience-producers exercised their experience-causing powers at all times.

But a simple adjustment to the thought-experiment undercuts Dainton's own account. We need only suppose that you are given an inescapable choice between (1) submitting to the bacon-slicer treatment one year from now, and (2) being put into a coma right now, to be revived a year later. It seems obvious that you should choose the second option (if you want to survive). Yet, if Dainton's

There is another problem with Dainton's account, connected with the fact that there does not need to be any temporal gap between two complete streams of consciousness belonging to the same mental life. When we think of divisions between the successive streams of consciousness that make up a conscious mental life, we tend to think of a stream that ends with the phenomenology that typically characterizes the experiences one has when falling asleep, and a new stream that begins with the phenomenology that typically characterizes waking up. And we commonly think of the falling-asleep experiences as occurring some hours before the waking-up experiences. But this temporal gap is unnecessary to keep the two streams of consciousness separate. There is no contradiction in the idea that my falling-asleep experiences are followed immediately by my waking-up experiences, with no interval of unconsciousness (for me) in between.

What can a hypothetical-stream theorist say about this kind of case? According to Dainton, the stream that ends with the falling-asleep phenomenology belongs to the same conscious self as the stream that begins with the waking-up phenomenology, only if the producers of the experiences in these streams would have produced experiences belonging to a *single*, extended stream of consciousness, had all potential producers of experience actually produced experiences at all times. But in the case we are considering, all (relevant) potential producers of experience *do* produce experiences at all times, and yet the experiences I have (falling asleep, waking up) do not belong to a single stream of consciousness. Dainton's theory therefore compels us to say that, in the case described, the self that falls asleep is different from the self that wakes up.[15]

theory is right, then you cannot survive either way. For if you opt for (2), it is false that what produces experiences in the revived subject would have produced experiences co-streamal with experiences produced by the brain of the pre-comatose subject, had everything with the potential to produce experience actually produced experiences over the year prior to revival.

[15] As was the case for the other two reductive analyses of co-subjectivity that we have considered, the hypothetical-stream analysis entails that co-subjectivity is a matter of degree. That is because it is a matter of degree whether there exists, at a given time, some unrealized potential for experience. At one end of the spectrum, there is the brain of someone who is in deep dreamless sleep; this is a clear case in which there is an unrealized potential for experience. At the other end of the spectrum, there is a jar of cremation ashes; this is a clear case in which there is no potential for experience. In the middle, we have the case in which scientists make a detailed map of your body (including your brain) and then euthanize you and dispose of your body. Does the existence of the map imply that there persists an unrealized potential for experience in this case? The answer might depend on whether the technology exists to construct a new body on the plan of the old one. In any event, it seems clear that the degree to which there exists an unrealized potential for experience in this case is higher than in the ashes case, and lower than in the dreamless sleep case. See Dainton (2008), 114–30, 311–40 for relevant discussion.

10.4 Particularism about the self

What should we conclude from the failure of the reductive accounts of the conscious self discussed above? One possibility is that there is a successful account that we have not yet considered. But it is very hard to see what form such an account might take. Another possibility is that co-subjectivity is a simple, unanalyzable relation; this is the primitivism that Leibniz favors. Yet another possibility is that there is no such thing as inter-stream co-subjectivity—that the whole idea of a conscious mental life comprising multiple streams of consciousness is somehow misguided; this is the anti-realist position.

But a less dramatic possibility is that different criteria guide our application of the "same self" concept on different occasions, so that there is no true and illuminating completion of the schema:

Several experiences belong to the same conscious self if and only if... [16]

On this particularist view, our concept of a conscious self does not have the kind of stability that it would need to have, in order to submit to the kind of analysis that those working in the reductive tradition have sought. Sometimes by "conscious self" I might mean something to do with a physical or functional substrate of experience. On other occasions, I might mean something related to memory. On still other occasions, I might mean something connected with possible streams of consciousness. And I might sometimes mean something that combines these factors in one way or another.

The concept of a conscious subject that encompasses multiple streams of consciousness is not exactly ambiguous on this view, but it is not univocal either. We mean different things by "same conscious subject" on different occasions, depending on the circumstances particular to those occasions, but it is no mere accident that we mean the various things we do. The phrase "same subject" (or "same self") is not like the phrase "same bank." That is why we may speak of *the concept* of co-subjectivity, even though it would be a mistake, according to particularism, to speak of *the relation* of co-subjectivity.

Is particularism about the conscious self compatible with radical phenomenalism? It is, provided that radical phenomenalism is compatible with each of the things we mean on different occasions by saying that one experience is

[16] Parfit argues that when it comes to the self, talk of identity is best replaced with talk of survival; see Parfit (1971), 4–10. Supposing he is right, the schema I claim to resist useful completion is: "The subject of an experience e_1 survives as the subject of an experience e_2 if and only if..." Of course, it may be necessary to replace the "if and only if" with "to the extent that," depending on the details of the offered analysis.

co-subjective with another. But on each occasion, what we mean by this is what one of the reductive theories of the self says we mean (or should mean) by it on *all* occasions. More cautiously, for every occasion *O* on which we predicate co-subjectivity of two experiences, there is a relation *R* such that:

(1) *R* is one of the relations to which reductive theories attempt to reduce co-subjectivity, or a relation that is a logical construction out of two or more of those relations; and[17]
(2) by predicating co-subjectivity of two experiences on occasion *O*, we mean that the experiences stand in *R*.

So, a particularist view of the conscious self is consistent with radical phenomenalism, provided that radical phenomenalism is consistent with each of the reductive analyses of the self considered earlier. And to show that radical phenomenalism is consistent with each analysis, it is enough to give a phenomenalist version of each analysis—a version that is compatible with an atemporal conception of experience.

The easiest way to see that phenomenalist versions of these analyses are possible is by calling on the phenomenalist account of embodiment developed in Chapter 8.

Begin with the substrate theory. According to this, experiences in different streams of consciousness belong to the same mental life to the extent that the experiences correspond to states of a single object, like a brain, that we can understand in purely non-mental terms. Standard versions of the substrate analysis think of this correspondence causally, as a matter of a single brain giving rise to experiences in different streams at different times. But we can also think of the correspondence non-causally. A substrate theorist could say that two experiences belong to the same conscious subject just in case they are embodied by states or processes of the same physically- or functionally-individuated brain, where embodiment is understood in terms of the non-causal relationship described in Chapter 8. According to a phenomenalist, this brain and its states and processes are phenomenalistic constructions, but that needn't prevent us from referring to them in a substrate analysis of the conscious self. A phenomenalist substrate theorist could think of selves as second-order metaphysical constructions, logically posterior to the constructions known as bodies.

Next, consider the asymmetric dependence theory. Here, too, we can use the phenomenalist account of embodiment to formulate an analysis that captures the basic idea behind the theory. A phenomenalist sympathetic to the asymmetric dependence theory could point out that a conscious memory-experience has

[17] One such logically constructed relation is the relation that two experiences bear to one another if, and only if, they arise from the same brain *or* stand in a suitable relation of asymmetric dependence.

a phenomenalistically constructed embodier (e.g. neural correlate), as does the experience of which it is a memory. He could then say that one experience, *e*, belongs to the same conscious mental life as another experience, *e′*, just in case (1) the neural correlate of one of the experiences in the stream to which *e* belongs asymmetrically depends on the neural correlate of one of the experiences in the stream to which *e′* belongs or (2) the neural correlate of one of the experiences in the stream to which *e* belongs asymmetrically depends on the neural correlate of one of the experiences in a stream, one of whose experiences' neural correlate asymmetrically depends on the neural correlate of one of the experiences in the stream to which *e′* belongs, or (3) . . . etc. By this account, it is only in a derivative sense that an experience in one stream of consciousness depends on an experience in another stream: the one experience depends on the other only in the sense that its embodier depends on the embodier of the other. The account nonetheless retains the spirit of the asymmetric dependence analysis, and it seems appropriate to consider it a phenomenalist version of that analysis.[18]

Finally, consider the hypothetical-stream theory. On Dainton's version of this theory, experiences in different streams are co-subjective just in case their respective causes would have caused experiences belonging to a single stream of consciousness, had all potential for consciousness been realized throughout the shortest period of time that includes both experiences. This account does not sit well with radical phenomenalism, which does not think of conscious experiences as having causes (or being in time). But we can use the phenomenalist account of embodiment to formulate a version of the hypothetical-stream theory that is consistent with a conception of conscious experiences as uncaused by physical events, and unlocated in time.

Take two experiences, e_1 and e_2, belonging to different streams of consciousness. A radical phenomenalist interested in adopting the hypothetical-stream analysis of selfhood could say that these experiences belong to the same conscious self, provided that the following condition is satisfied: if every phenomenalistically-constructed brain existing at some phenomenalistically-constructed time between the neural correlate of e_1 (call it $N(e_1)$, occurring at t_1) and the neural correlate of e_2 (call it $N(e_2)$, occurring at t_2) had, at each moment of its existence, been physiologically awake and active, then the brain containing $N(e_1)$ would have contained, at t_1, a neural correlate of an experience e_1', and the brain containing

18 Alternatively, a phenomenalist might posit relations of direct, non-derivate dependence between various experiences in different streams. That would also yield a version of the asymmetric dependence theory that was compatible with radical phenomenalism.

$N(e_2)$ would have contained, at t_2, a neural correlate of an experience e_2', such that e_1' and e_2' would have belonged to a single stream of consciousness.

Taken as attempts to specify what we always mean (or should mean) by saying that two experiences belong to the same conscious self, the phenomenalist versions of the reductive accounts of conscious selfhood are no better than the original, non-phenomenalist versions. But taken as specifications of various things that we mean by saying (on various occasions) that experiences occur in the same conscious mental life, the phenomenalist versions of the substrate, asymmetric dependence, and hypothetical-stream theories are just as good as the original versions of those theories. By saying that two experiences belong to the same conscious mental life, we mean sometimes one thing, sometimes another, but never anything incompatible with an atemporal view of experience, or with the phenomenalist metaphysics defended in this book.

10.5 Conclusion

In the Western tradition, most anti-reductionists about the conscious self are also primitivists about it; in Indian Buddhist philosophy, the most popular alternative to reductionism about the self is anti-realism. Almost all authors, Eastern and Western, tend to assume that primitivism and anti-realism are the only alternatives to reductionism, at least when it comes to the self. Russell supplies an unusually explicit manifestation of this tendency:[19]

> So you can collect a whole set of experiences into one string as all belonging to you, and similarly other people's experiences can be collected together as all belonging to them by relations that actually are observable and without assuming the existence of the persistent ego [that is, without taking a primitivist view of the conscious self]. It does not matter in the least . . . what exactly is the given empirical relation between two experiences that makes us say, "These are two experiences of the same person." It does not matter precisely what that relation is, because . . . the mere fact that you can know that two experiences belong to the same person proves that there is such an empirical relation to be ascertained by analysis. Let us call the relation *R*. We shall say that when two experiences have to each other the relation *R*, then they are said to be experiences of the same person. That is a definition of what I mean by "experiences of the same person."[20]

[19] In debates about personal identity, non-reductionism has come to be known as "anti-criterialism"; see Merricks (1998), 116. Recent exponents of primitivist anti-criterialism include Chisholm (1976), 104–13, Swinburne (1984), 17–26, and Lowe (1996), 39–44. For the Buddhist tradition, see, for example, Garfield (1995), 48–9, Edelglass and Garfield (2009), 269, and the discussion in Chapter 4 of Garfield (forthcoming).

[20] Russell (1918), 149–50.

Here Russell argues that the possibility of a reductive analysis of the conscious subject (or "person") is guaranteed by the fact that we often know that two experiences belong to the same subject by knowing that the experiences stand to one another in some empirical relation. This inference, from:

Whenever two experiences belong to the same conscious self, it is by standing to one another in some empirical relation,

to:

There is some empirical relation whose obtainment between any two experiences is necessary and sufficient for those experiences to belong to the same conscious self,

can be tempting—it obviously seduced Russell—but it is a temptation that we should resist. A failure to notice that the inference is invalid is sure to go along with a failure to recognize particularism as an alternative to reductionism, primitivism, and anti-realism.[21]

Ideally, one would like to have a defensible reductive analysis of the conscious self. Failing that, I have argued that we ought to favor particularism about the self. Particularism allows us to recognize the existence of conscious selves, or at least the truth of various positive judgments in which we apply the concept of a conscious self, without adding selves to our stock of ontic fundamentals.

[21] Even in philosophical debates in which particularism is recognized as a live option, the debaters frequently lose sight of it; see Cargile (1998), 247, 260–1, 275.

11

Intentionality

The stream of conscious thoughts, memories and perceptions seems to have a life of its own that is constituted independently of its external environment. This is intuitively supported by an obvious thought experiment. Apparently I can imagine what it is like to be an isolated brain that is a physical duplicate of my own brain. What I imagine includes not just that brain's non-intentional phenomenal states, its flutters and pains, but also states and events that correspond to my own outward directed thoughts and perceptions. I imagine my isolated twin's states and events as subjectively representing things in the same manner as those thoughts and perceptions of mine. The intuition supports the view that my own mental stream's intentional features—even those of its outward directed thoughts—are constituted independently of my actual situation in the world.

Brian Loar, *Phenomenal Intentionality as the Basis of Mental Content*

11.1 Introduction

Thought is the first thing that metaphysical inquiry demands of us, but it may be the last thing whose nature metaphysical inquiry reveals. So it is in the present case. Until now, I have said very little about mental representation, and offered nothing by way of a philosophical account of it.

There is a certain view which, if correct, would free the phenomenalist from any requirement to give an account of mental representation or (as philosophers like to call it) "intentionality." This is the view that conscious experiences are, by their very nature, representational. In this view, intentionality is automatically given with phenomenology, so that there is no need for a theory, such as phenomenalism, which takes phenomenology as ontologically basic, to offer a separate account of intentionality.[1]

[1] The claim (1) that conscious experience is essentially intentional does not entail either (2) that facts about conscious experience reduce to facts about intentionality, or (3) that facts about intentionality reduce to facts about conscious experience. Nor does (3) entail (1) (you could maintain that some, but not all, conscious experience is essentially intentional; for example, that intentionality is an

But the view that phenomenology is essentially intentional is contentious. Some hold that the intrinsic nature of a conscious experience is exhausted by its what-it-is-likeness or "raw feel," where this feel needn't have any representational content. This is the view to which I myself am inclined.

My purpose in this chapter is to show that even if, as I suspect, conscious experiences are not essentially representational, a radical phenomenalist can reduce spacetime and its contents to purely phenomenological facts, without dragging intentionality into the picture as an additional term of analysis.

Some deny the possibility of *any* reductive analysis of intentionality. According to them, the fact that we can represent the world as being certain ways is just a primitive fact, and intentional states (of believing, desiring, etc.) are best left as primitive elements of our metaphysical theories. Others are more optimistic about the prospects for understanding intentionality in terms of something more basic.[2]

I side with the optimists.

Now, it may seem that I have less ground for optimism here than anyone. Radical phenomenalism analyzes physical states of affairs partly in intentional terms. For a totality of experiences to be interpretable as satisfying such-and-such conditions is for it to be the case that certain *beliefs* about that totality would be reasonable beliefs to have. The notion of interpretation that the phenomenalist analysis invokes is an essentially intentional notion.

A successful phenomenalist analysis of the physical must invoke the concept of a possible belief, and therefore, implicitly, the concept of intentionality. A successful phenomenalist analysis of *intentionality* must therefore *not* invoke the concept of anything physical.[3]

At first, this might look like an intolerably severe constraint on any attempt to provide a reductive account of intentionality. All of the best-known theories of mental representation deploy physical, or at least spatiotemporal, concepts in one

essential feature of perceptual phenomenology, but not of bodily sensations like dizziness and nausea). For (1), see Horgan and Tienson (2002). For (2), see Kriegel (2003). (3) is the position defended in this chapter, and, in other forms, by Russell (1910–11), Strawson (1994), Loar (2003), and Farkas (2008).

[2] For intentional primitivism, see Brentano (1874/1995), 68. For reasons to prefer reductionism, see Field (1978), Dretske (1981), xi, and Fodor (1987), 97.

[3] The logical situation here is somewhat subtle, actually. There is nothing to prevent a phenomenalist from saying that facts about intentionality *supervene* on physical facts. Maybe they do—in which case, they also supervene on phenomenological facts (namely, the phenomenological facts on which the relevant physical facts supervene). The point being made here is that the phenomenalist must be able to understand any such supervenience in terms of a deeper analysis of the intentional in terms of the phenomenological. Absent such an analysis, an appeal to the (supposed) reducibility of intentionality to something purely physical would render circular the phenomenalist analysis of physical states of affairs.

way or another. Worse yet for a phenomenalist who seeks a reductive analysis of intentionality, the majority opinion among philosophers today is that it is demonstrably impossible to analyze intentional states without reference to features of the physical environments of those whose intentional states they are.

But no phenomenalist worth his salt will be cowed by this near-consensus. Instead, he will attempt to articulate a *purely phenomenological* conception of intentionality. The purpose of this chapter is to show that such a conception is both possible and defensible. I am not going to try to achieve this by providing detailed analyses of particular intentional states. Rather, my goal is to give a convincing "proof of concept"—to show that there is nothing in principle to prevent us from providing such analyses. That will be enough to clear phenomenalism of any suspicion that it takes intentionality as a metaphysical primitive.

11.2 Phenomenalistic intentionality

In a well-known discussion about the nature of representation, Hilary Putnam asks us to consider the case of a monolingual English speaker who, under the influence of hypnosis, visually scans a Japanese newspaper with the feeling of understanding the words he is scanning. We are to suppose that the man's stream of consciousness while scanning the paper is phenomenally indistinguishable from that of an ordinary Japanese person reading the same paper (as Putnam puts it, the similarity is "so perfect that the person could even fool a Japanese telepath"). Putnam contends that despite all of this, if the reader "couldn't use the words in the right contexts, answer questions about what he "thought," etc., then he didn't understand them."[4]

Putnam is surely right about this. But suppose that we draw out the endpoints of the hypnotic episode, so that it covers the man's entire life. Psychologists have somehow interfered with his mind and controlled his environment so as to make it seem to him that he understands Japanese whenever he encounters it in writing or in speech. They have arranged things so that his entire life is phenomenally indistinguishable from that of some ordinary literate Japanese person.

Still, one might argue, even in this case, the man does not understand Japanese, assuming that, as in Putnam's example, he would become flummoxed by any Japanese questions besides those that are actually put to him, would find himself at a loss confronted with any Japanese texts besides those that he actually encounters, etc.

[4] Putnam (1982), 2.

But now suppose that the psychologists have arranged things so that the man is indistinguishable from an ordinary Japanese person not only with respect to his actual phenomenal states, but also in terms of his phenomenal dispositions. Statements of the form, *Had his conscious mental life possessed phenomenal feature* ϕ*, it would also have possessed phenomenal feature* ϕ' are true for all values of ϕ and ϕ', for both the hypnotized subject and his ordinary Japanese counterpart.

In this case, I submit that the hypnotized man *does* understand Japanese, and that he is in fact indistinguishable from his Japanese counterpart in all intentional respects.

The idea of someone hypnotized from cradle to grave is hard to grasp. Let's change the example, so that instead of talking about a hypnotized subject, we are talking about a Leibnizian monad. This monad, we are to imagine, has, over the course of its existence, the same phenomenal states and dispositions as the Japanese man, whom I'll call Mr Nakamura. Nothing *causes* the monad to have these states or dispositions: the dispositions are just part of its intrinsic nature, and its conscious mental states (which serve as the inputs and outputs of its dispositions) occur in it without being *made* to occur in it by anything external to the monad. As Leibniz would have put it: the monad "has no windows" (nor, he might have added, any doors, radios, telephones, or internet connection).

According to Putnam, the monad has no intentionality—no beliefs, no desires, no perceptual experiences, etc. This, however, is highly implausible. Maybe I have good reasons to believe that I am not a monad, but the fact that I believe that grass is green is not one of them.

We cannot conceive of two beings that duplicate one another in terms of their conscious experiences and phenomenal dispositions, but who differ in some intentional respect (e.g. by one of them but not the other believing that grass is green, or one of them but not the other desiring that it stop raining). And if we cannot conceive of two beings duplicating one another with respect to *A* while differing from one another with respect to *B*, we have every reason to believe that how a thing is with respect to *A* completely determines how it is with respect to *B*. So, we have every reason to believe that a being's phenomenological features completely determine its intentional features.[5]

That is the basic argument for internalism about mental representation. Like all conceivability arguments, it is subject to the complaint that what we can or cannot conceive of is not an infallible guide to what is or isn't possible. But even if inconceivability does not guarantee impossibility, we ought to be skeptical about the possibility of any state of affairs that is prima facie impossible, defies coherent

[5] I take this to be the central argument of Loar (2003), which is the source of this chapter's epigraph.

description, and resists all efforts to demystify. And that means being skeptical about the suggestion that beings with identical phenomenal dispositions could differ in the representational contents of their mental states.

I have suggested that the intentional features of a mind supervene upon its phenomenological features (conscious mental states and phenomenal dispositions). However, as mentioned earlier, the default position in contemporary philosophy is that a person's intentional features, far from supervening upon his phenomenological features, do not supervene on his intrinsic features at all, even when these are taken to include the person's intrinsic *non*-phenomenological features. The reason why this is the default position is that there are a couple of seemingly powerful arguments against the view that one's intentional states supervene upon one's intrinsic features. Let's scrutinize these arguments now.[6]

11.3 Internalism versus externalism

In this section, we consider two influential arguments against the idea that a being's intentional states supervene on its intrinsic features (such as its intrinsic phenomenological features). These arguments, and the debate surrounding them, have not taken place against the background of a phenomenalist metaphysics, and to recast them into phenomenalistically acceptable language would be both tedious and pointless. My objective is to show that the arguments do not work, and it will be enough for this purpose if I can show that they do not work even on the anti-phenomenalist assumptions upon which the arguments' proponents predicate them. So, from this point until the end of the section, we will suppress our phenomenalist scruples, and conduct the discussion in more conventional terms.

The debate between internalists and externalists concerns the relationship between an individual's intentional states and his intrinsic properties. A thing has a property F *intrinsically* just in case it has F, and could have F even if it were the only thing. (More precisely: a thing with property F has the property intrinsically just in case it could have F, even if it were the only contingently existing thing in existence: nothing can have any property in the absence of necessarily existing things, if there are such things.) A thing's intrinsic properties are just the properties it has intrinsically.

The usual way to frame the debate between internalists and externalists is with the question whether two people who had all their intrinsic properties in common

[6] Some of what I say in the following section is anticipated by Pelczar (2009).

could differ with respect to the contents of their mental states. Externalists supposedly say yes, internalists no.

This is a bad way to frame the debate. It is a bad way to frame it because *it is impossible for two individuals to share all of their intrinsic properties.*

For each of your parts, P, you have the intrinsic (to yourself) property of containing P. For any way that two of your parts relate to one another, it is also an intrinsic property of yourself that you contain parts related to one another in that way. So, anything that had all of your intrinsic properties would contain all of your parts related to one another as they are related in you. But any such thing would just *be* you, and hence would not differ from you in the contents of its mental states.[7]

The usual way of framing the debate therefore begs the question against externalism. How then should we frame it?

Not by confining our attention to non-relational intrinsic properties. (F is a *relational* property if and only if there is some individual such that that individual must exist in order for anything to have F. An example is the property of being a French resident, which requires the existence of France for its instantiation.) It is true that this restriction would prevent internalism from becoming a trivial truth, by taking out of consideration properties like that of containing P (where P is one of one's own parts), but it is otherwise completely arbitrary. What is distinctive about externalism is its implication that the contents of a person's mental states depend not just on those of his features which he could have even if he were the only being in existence, but also on facts about his natural or social environment. If we were to define externalism specifically in terms of non-relational intrinsic properties, it would not carry this implication; the claim that two individuals might differ in the contents of their mental states despite sharing all their *non-relational* intrinsic properties is compatible with the claim that the aforementioned environmental factors are completely irrelevant to the contents of a person's thoughts. An internalist could maintain that what must be added to a person's non-relational intrinsic properties in order to determine the contents of his thoughts are simply his *relational* intrinsic properties, such as the property of containing such-and-such parts related to one another in such-and-such ways.

The correct way to frame the disagreement between internalists and externalists is with the following question:

[7] Here I omit from consideration relations that two of one's parts bear to each another only by virtue of bearing some other relation to a third thing existing outside of oneself; an example would be the relation designated by the following formula: "*x* is adjacent to *y* and *y* lies within a light year of the Sun" (*x* and *y* might be a couple of your brain cells).

Given that two people differ with respect to the contents of their mental states, must this difference reduce to a difference in their intrinsic properties?

Internalists say yes, externalists say no.

This way of framing the debate does not prejudge its outcome one way or the other. In particular, it does not make internalism come out trivially true. For, from the fact that even doppelgangers must differ in some of their intrinsic properties (namely their relational intrinsic properties), it does not obviously follow that every difference in the contents of two doppelgangers' mental states reduces to a difference in the doppelgangers' intrinsic properties.

To work, an argument against internalism must show that there can be differences in the contents of doppelgangers' mental states that do not reduce to differences in the doppelgangers' intrinsic properties. To defend internalism against such an argument, we must either show that the supposed differences in content are illusory, or else grant the differences, and show that for each such difference there is a difference in the doppelgangers' intrinsic features to which the granted difference in content reduces.

Arguments against internalism take two basic forms. Arguments of the first form focus on intentional states that concern general categories or natural kinds, such as chemical substances (aluminum, molybdenum, water) and biological species (tigers, beeches, elms). These arguments, reviewed below, succumb to an objection due to Tim Crane. This is an objection to the effect that the supposed differences in content between the doppelgangers' mental states are illusory.

The second kind of argument against internalism focuses on intentional states that we typically express using indexical terms, such as "I," "here," "now," and "this." Although widely regarded as decisive, this kind of argument is, I argue, also subject to refutation, by means of an objection that grants the supposed differences in doppelgangers' mental contents, but argues that these differences reduce to differences in the doppelgangers' intrinsic properties.

11.3.1 Natural kinds arguments

Twin Earth is familiar territory to philosophers: a planet just like Earth, except that wherever there is H_2O on Earth, on Twin Earth there is a liquid that has all the outward qualities of H_2O, but a different molecular composition, call it XYZ. In 1750, no one on Earth or Twin Earth knew the molecular composition of the stuff in their respective lakes, wells, etc. So, consider Oscar and Twin Oscar ("Toscar," for short). We are to imagine that Oscar and Toscar are "doppelgangers": perfect physical and phenomenal duplicates of one another (except for the fact that Oscar is about 60 percent H_2O, whereas Toscar is about 60 percent XYZ—a difference we may ignore as irrelevant). Also, Oscar and Toscar have the same quality of conscious experience: when Toscar drinks a cold glass of water, he has the same thirst-quenching sensation as Oscar, etc. Whatever utterances Oscar is prepared to make using the word "water," Toscar is prepared to make as well. Oscar utters things like: "Water boils at a hundred degrees Centigrade and freezes at zero degrees," "Liquid water is denser than solid water," "Clouds consist of water vapor," and "I would like a glass of water." Toscar makes exactly the same utterances.[8]

But when Oscar says "Please bring me some water," he gets H_2O, whereas Toscar gets XYZ when he utters the same words. Similarly, if Oscar neglects to pay the bills that have "Water Bill" printed on them, the result is that H_2O stops flowing from his tap, whereas if Toscar neglects to pay bills with the same heading, the result is that XYZ stops flowing. And so forth. Externalists think that such considerations show that when Oscar sincerely says "Water is wet," he is in a belief state that is true if and only if H_2O is wet, whereas when Toscar utters the same words sincerely, he is in a belief state that is true if and only if XYZ is wet.

Or take the case of Hilary and Thilary, who, like Oscar and Toscar, duplicate one another molecule-for-molecule (and, at the level of conscious experience, quale-for-quale). Neither Hilary nor Thilary can distinguish beech trees from elm trees, and they both have the same vague mental image when asked to imagine a beech or an elm. But on Twin Earth, the words "beech" and "elm" are switched, so that "beech" is the accepted term for elms, and "elm" the accepted term for beeches. So when Hilary goes to the nursery and says "please give me a dozen beech trees," what he gets is a dozen beeches, whereas when Thilary says "please give me a dozen beech trees," what he gets is a dozen elms; etc.

Now suppose someone says to Hilary: "elms grow taller than beeches." If Hilary accepts this person's statement, he is, according to externalists, in a belief state that is true if and only if elms grow taller than beeches. But when Thilary accepts some Twin Earthling's statement "elms grow taller than beeches," he is, according to externalists, in a belief state that is true if and only if beeches grow taller than elms. Thus, Hilary and Thilary have intentional states with different contents, despite duplicating each other in all relevant intrinsic respects.

As they stand, these arguments against internalism simply do not work. For, as Crane points out, the cases described above are consistent with the view that Oscar and Toscar (or Hilary and Thilary) are completely alike in the contents of their intentional states. If, for example, Hilary and Thilary both use "elm" as a term to cover both beeches and elms, their uses of this term still have the different

[8] For the origin of this thought-experiment, and the natural kinds arguments based on it, see Putnam (1973), 700–4; see also Burge (1977).

results described earlier (Hilary would bring home elms from the nursery, whereas Thilary would bring home beeches, etc.). Similarly for Oscar and Toscar: their uses of "water" will have the different consequences described earlier—Oscar's waiter will bring him H_2O, Toscar's will bring him XYZ, and so on—even if by "water" they both just mean drinkable stuff that freezes at zero Centigrade, boils at a hundred Centigrade, and is denser as a liquid than as a solid. Since nothing about these thought-experiments forces us to attribute to Oscar and Toscar beliefs with different truth conditions (or desires with different satisfaction conditions), they simply fail to call internalism in question.[9]

Suppose Oscar were to learn that the stuff he drinks is H_2O and the stuff they drink on Twin Earth is XYZ. Would he then say that all along, by "water" he had meant H_2O? If so, then Oscar, as he actually is (knowing nothing about chemistry or Twin Earth), has a certain disposition (of which he is not aware) to say that by "water" he has always meant H_2O; if not, then Oscar lacks this disposition. Likewise, Toscar either does or does not have the disposition to answer "Yes" if asked whether he has always meant XYZ by "water," upon learning that the stuff he drinks is XYZ (rather than H_2O).

More generally, each man is disposed to make certain judgments about what he means, and has meant, by "water," upon receiving relevant information. Call these *"water"-dispositions.*

Two cases are possible.

First, it could be that Oscar and Toscar have identical "water"-dispositions. In this case, they simply mean the same thing by "water." After all, what better reason could we have for saying that Oscar and Toscar mean the same thing by "water"? A speaker's considered judgments about what he means by his words are surely our best guide to what he means by them, especially when those judgments are made in the light of all relevant information. If Oscar and Toscar are disposed to identify the same stuff (or class of stuffs) as water in the light of all relevant information, who are we to insist that they actually mean something different by their respective uses of the word "water"? Presumably *their* fully-informed opinion about what they mean counts for more than *our* opinion about what they mean; certainly it should not count for *less*.

Second, it could be that Oscar and Toscar differ in their "water"-dispositions. In this case, they might mean something different by "water." But if Oscar and Toscar differ in their "water"-dispositions, then they differ in some intrinsic respect—some respect by virtue of which they are disposed to make different judgments about what they mean (or have meant) by "water," given all relevant

[9] For Crane's discussion, see Crane (1991), 10–13.

information. So in this case, Oscar and Toscar are not, as Putnam's argument requires, doppelgangers.[10]

Since Oscar and Toscar mean the same thing by "water" in every possible scenario in which they are doppelgangers, no possible scenario confirms the externalist claim that an agent's intrinsic features underdetermine the contents of his judgments about water (or elms, or whatever).

The influence of the natural kinds arguments may be due partly to a failure to distinguish clearly between (1) what people mean by the words they use, and (2) what the words themselves mean. What the word "water" means in a given linguistic community is a function of what those who use it in that community mean by it (although not a straightforward function—democracy does not always rule). Unless the linguistic community consisted of just one person, it would be absurd to hold that the meaning of a word supervened on the intrinsic properties of any single speaker. But it would be a mistake to infer from this that what a given person *means by* a word does not supervene on that person's intrinsic properties. At most, the natural kinds arguments show (what is anyway obvious) that what the words a person uses mean does not depend only on that person's intrinsic properties. They do not show that what a person means by the words he uses depends on something beyond his intrinsic properties; much less do they show that the contents of his mental states depend on more than this.[11]

11.3.2 The indexical argument

Natural kinds arguments fail to refute internalism because they give us no compelling reason to believe that doppelgangers' mental states differ in their truth or satisfaction conditions. The situation is different, however, if, following Putnam, we suppose that natural kind terms like "water" function as indexicals. Imagine that Oscar lives at the dawn of modern chemistry, so that while the concept of a molecule is available, it has not yet been discovered what kind of molecules make up the Earthly substance known as "water." Now suppose that Oscar defines the word "water" for himself as follows: "By 'water,' I mean whatever consists of molecules that have the structure exhibited by the molecules that fill this glass"—raising the glass that he is holding. It is hard to deny that when Oscar proceeds to use the word "water" in thought and speech in accordance with this definition, he uses it to form and express beliefs and desires whose truth or satisfaction

[10] Alternatively, Oscar's "water"-dispositions might be a primitive, ungrounded feature of Oscar, and *mutatis mutandis* for Toscar. In this case, the two men differ in their ungrounded dispositions, which are equally intrinsic to them as grounded dispositions.

[11] The speaker-meaning/word-meaning distinction does lead a sort of incognito existence in these debates, in the form of a distinction between "narrow content" and "wide content."

conditions have to do specifically with H_2O rather than XYZ. For example, if Oscar thinks to himself: "There is water on the floor," he is in a mental state that is true if and only if there is H_2O on the floor. Supposing that Toscar has defined "water" in his idiolect in a similar manner, when he thinks to himself "There's water on the floor," his thought is true if and only if there is XYZ on the floor.[12]

Let's call this *the indexical argument,* since it focuses on beliefs of the sort that we typically express using indexical language. Putnam's supposition that natural kind terms function as indexicals is incidental to the argument; he could simply point out that when Oscar says "I'm hungry!" he means something that is true if and only if Oscar is hungry, whereas when Toscar utters the same words sincerely, he means something that is true if and only if Toscar is hungry. So the truth conditions of their utterances differ, as do the contents of their underlying belief and desire states: Oscar is in a belief state that is true if and only if Oscar is hungry, and in a state of desire that is satisfied if and only if Oscar eats; *mutatis mutandis* for Toscar.

On the face of it, this is a strong argument against any thoroughgoing internalism about mental representation. Unlike the beliefs that the doppelgangers express with the sentences "Water is wet" and "Elms make great firewood," it is hard to argue that the belief Oscar expresses with the words "I am hungry" has the same truth conditions as the belief Toscar expresses with these words.

But this appearance of strength quickly evaporates when we reflect that a person's success at referring to himself does not—or at any rate need not—depend on anything besides his intrinsic properties. If, on some occasion, Oscar succeeds in referring to himself, for instance in the course of thinking to himself, *I am hungry*, or, *I call the liquid in the glass I am holding "water,"* then, intuitively, he refers to himself in any possible situation in which he has the same intrinsic properties as he has on this occasion. Certainly, his success at self-reference does not depend on anything extrinsic to him, since he does not differ from Toscar in any relevant extrinsic respect; Oscar's success in referring to himself has nothing to do with the fact that his environment contains H_2O rather than XYZ.

Oscar succeeds in representing himself. That much is certain, and anyway required by the indexical argument against internalism. He does not succeed in doing so by possessing some extrinsic property that Toscar lacks, since, for the purposes of self-reference, Oscar and Toscar are alike in their extrinsic properties. Nor does Oscar succeed in representing himself by dint of his non-indexical beliefs, since these do not differ in content from Toscar's non-indexical beliefs (for the reasons discussed earlier), and even if we were to grant the differences

[12] For Putnam's discussion, see Putnam (1973), 706–10.

in non-indexical belief content that Putnam posits, such differences would not account for the fact that Oscar refers to himself rather than Toscar in such thoughts as *I call the liquid in the glass I am holding "water."*

That leaves just two options. Either it is a brute, irreducible fact about Oscar that he represents himself in thought and speech (and likewise for Toscar), or else Oscar represents himself by virtue of possessing certain *intrinsic* properties.

To say that it is a brute fact that Oscar represents himself is just to say that it is a brute fact that he has the property of representing himself, and that is just to say that his having this property does not reduce to his having any other property, intrinsic or extrinsic. But if Oscar has a property that does not reduce to his having some other properties, intrinsic or extrinsic, then he has that property intrinsically. For if he had it extrinsically, his having it would reduce to his relating in a certain way to his environment.

The upshot is that Oscar's success in representing himself either reduces to his having certain intrinsic properties, or else just *is* a matter of his having the intrinsic property of representing himself. Either way, an internalist account of self-representation turns out to be the correct one.

Any thought that you can express with a sentence of the form, "I am such that . . ." is a thought that you can also express with a sentence of the form, "The subject of *this* mental state is such that . . ."—where "*this*" picks out one of one's own mental states. Likewise, any thought expressible with a sentence of the form, "It is here such that . . ." is expressible with a sentence of the form, "The location of the embodier of *this* mental state is such that . . .", and any thought expressible with a sentence of the form, "It is now such that . . ." is expressible with a sentence of the form, "The time at which the embodier of *this* mental state occurs is such that . . .". To establish the reducibility of the thoughts that we express using indexical terms to intrinsic features of those who have the thoughts, it is therefore enough to establish the reducibility to such features of thoughts we express using the indexical "this."

Suppose that Oscar drinks a cold glass of water on a hot day, and says to himself: "This is a pleasant experience." How does Oscar succeed in referring to his experience? Well, however he manages it, it is not by relying on some feature of his environment—not, that is, by having some extrinsic property. Oscar could refer to the refreshing experience, even if he were the only being in the universe. Likewise for Toscar and his refreshing experience. Oscar's and Toscar's beliefs have different truth conditions because Oscar and Toscar refer to different mental states. But the fact that they refer to different mental states is explained by differences in Oscar's and Toscar's *intrinsic* features: the features by virtue of which Oscar refers to one of his own mental states, and Toscar to one of his.

Since an internalist theory can account for the differences in what doppelgangers mean by "this," it can account for the fact that the belief that Oscar expresses by saying, "It's hot here now" differs in its truth-conditions from the belief that Toscar expresses with the same sentence, since, as noted above, those beliefs are also expressible by saying, "It's hot where and when the embodier of *this* experience occurs," where each doppelganger uses "this" to refer one of his own experiences.

11.4 Conclusion

In the view of intentionality defended in this chapter, an intentional state, such as a belief, reduces to some purely phenomenological—and, on the atemporal conception of experience, aspatiotemporal—state of affairs. But, like conscious experiences themselves, beliefs are (at least typically) embodied, in the manner described in Chapter 8. So even if it is strictly incorrect to conceive of a belief as a functional state or physical process spread out in time and space, such a conception does correctly capture the nature of that which *embodies* a belief (the belief's spatiotemporal correlate); likewise for desires and other intentional states.

Any viable idealism must invoke certain intentional concepts. Some idealists, like Leibniz, give these concepts, or at least intentionality itself, a foundational place in their metaphysics. In this chapter, I have taken a different approach, arguing that facts about intentionality, like facts about the physical world, reduce to facts about a phenomenological level of reality that transcends time and space.

The worst case scenario for the phenomenalist is that a phenomenalistic reduction of intentionality turns out to be untenable, the arguments of this chapter notwithstanding. That would not compel the phenomenalist to abandon his position, but it would require him to follow Leibniz by treating intentionality as a conceptual primitive. On the other hand, assuming that the foregoing defense of a phenomenological internalism about mental representation is successful, we can be phenomenalists without having to accept intentionality as a fundamental part of our metaphysics. We can hold to the view that the primitive stuff of reality is consciousness, with no admixture of cognition.

12

Conclusion

> I think that almost all serious fly rodders will agree that the fewer moving parts a reel has the better. One of the greatest compliments I've heard about a current model came from a competitor who grumbled, "Hell, that reel doesn't have anything in it."
>
> Lefty Kreh, *Fly Fishing in Salt Water*

> It is vain to do with more what can be done with less.
>
> William of Ockham, *Summa Logicae*

Each of us has experiences as of physical things, and it is possible to interpret all our experiences taken together as experiences of a single world of physical things occupying a common time and space, and standing to one another in various spatial and temporal relations. In Leibniz's famous image, the totality of experiences is like a collection of sketches that we can interpret as different perspective drawings of a single landscape.[1]

It did not have to be this way. It could have been that none of us had conscious experiences at all. It could have been that we, or some other beings, had conscious experiences that, unlike those we actually have, did not present any subjective appearance of things in time or space. It could have been that each of us had experiences interpretable as perceptions of a physical world, but that it was impossible to interpret our experiences collectively as perceptions of a single, unified physical reality. It could have been that my experiences were interpretable as perceptions of one world, yours interpretable only as perceptions of a completely different world.

All of these are ways things could have been, but none is the way things are. Our experiences are, in fact, as of things arranged in time and space, and do collectively

[1] Leibniz (1712/1989), 199.

admit of construal as perceptions of parts of a unique physical reality. They are, as we might say, *worldlike*.[2]

That this is so should, I think, strike us as remarkable, and, upon reflection, profoundly mysterious. But that is not how we ordinarily think about it. Ordinarily, we refer the worldlike quality of our experiences to the fact that we all experience the same world, encounter objects in a common space, and witness events in a common time.

I have argued that we should abandon this way of thinking. The physical world that we inhabit does not *explain* the worldlike quality of our experiences: it *is* the worldlike quality of our experiences, or rather it is the tendency for experiences to constitute a worldlike totality of the kind that our experiences do, in fact, constitute.

There was a time when idealism was a respected and widely debated metaphysic. Today, it is most notable for its utter absence from the intellectual world stage. What happened? It is true that idealism faces serious objections, some of them never dealt with before in a satisfactory way. But this can be said of any metaphysical theory, and the most important objections to idealism were known and discussed long before its abrupt fall from grace around the beginning of the twentieth century.[3]

One possible explanation for the decline of idealism is that latter-day proponents of the theory tended to motivate it by its supposed epistemological virtues, rather than the metaphysical virtues that attracted Leibniz and Mill to the idealist position. Idealists too often tried to sell their position as an antidote to external-world skepticism.

This was a mistake, for two reasons.

First, there is no real problem of external-world skepticism. We all believe in an external world, and believe ourselves to be justified in doing so. If someone expresses a sincere doubt about whether he has a body, that is not an invitation to philosophical conversation: it is a cry for help.[4]

Second, even if it were possible to cast doubt on our ordinary beliefs about the external world, this would be no less possible in an idealist scenario than in

[2] Or, as John Foster says, "world-suggestive" or "amenable to physical interpretation": see Foster (2008), vii, 108–15.

[3] Berkeleyan idealism has fared somewhat better than other idealist theories, with recent defenses in Robinson (1994) and Foster (2008).

[4] See Cargile (1997), 74. The point actually goes back to Descartes, which is somewhat ironic in the light of subsequent developments; see Descartes (1641/1984), 11, where Descartes writes: "The great benefit of these arguments is not, in my view, that they prove what they establish—namely that there really is a world, and that human beings have bodies and so on—since no sane person has ever seriously doubted these things."

any other. Idealism, like any viable metaphysics of appearance and reality, must recognize a distinction between veridical experiences and deceptive experiences. A skeptical hypothesis for idealism is just a hypothetical scenario in which most or all of our experiences fall into the latter category. In this respect, idealism is no different from any other metaphysical theory. It is notable that neither Leibniz nor Mill proposes his metaphysics as a response to an imagined skeptical challenge. Rather, they come to the idealist (and in Mill's case, phenomenalist) point of view along a path of relentless conceptual simplification: skepticism has nothing to do with it.[5]

I have argued for radical phenomenalism on its merits, rather than on the demerits of more popular world-views. But it does the phenomenalist no harm, and may do him some good, to point out the shortcomings of the naive picture of reality that stands behind the popular alternatives to his position.

In this picture, physical things have natures exhausted by the qualities that Locke identifies as fundamental (shape, size, and solidity), and stand to one another in fundamental relations of spatial distance and temporal order. Consciousness is sketched in as a feature of various living things. Space and time themselves are an invisible, implied background that provides the setting for all mental and physical happenings.

It used to be that this picture cohered well with a scientific image of the world. The atoms of Democritus, the light-waves of Huygens, even Boscovich's infinitesimal centers of force—we can understand all of these as miniature versions of familiar everyday things, like grains of sand, or ripples of water, or macroscopic sources of attraction and repulsion. The naive picture appeals to us because it *is* a picture: we can easily imagine particles arranged in geometrical patterns, or waves propagating through an ethereal medium, or pushes and pulls within an invisible force-field. But in the modern scientific image of the world, the spatial and temporal properties of things are not, except in special limiting cases, fundamental, or even clearly real. Indeed, from a modern point of view, many of Locke's "secondary" qualities take priority over the qualities that Locke identifies as primary. A puddle's temperature, for example, has a greater claim to reality than its shape or duration.[6]

The features of the naive picture that appeal to us are precisely those that make it naive. Modern science describes a physical world that is a collection of entities

[5] For Leibniz's and Mill's indifference to skepticism, see Popkin (2003), 261–73, and Skorupski (1994), 99.

[6] The theories of Democritus and Huygens are relatively well known, that of Boscovich less so, despite having been quite influential at one time: see, for instance, the numerous references to Boscovich (1763/1922) in Thomson (1904).

structured by the deeply counter-intuitive geometry of relativistic spacetime (or by the equally counter-intuitive algebra of quantum mechanics).[7] In this world, ordinary objects exist only as challenging constructions. The only question is: constructions out of what? Rejecting the phenomenalist's answer to this question will not make the question go away, and alternative answers will not easily satisfy those who have eaten from the Tree of Science.[8]

Suppose we visualize the sum-total of humanly attainable knowledge as a geometric solid. The outer surface of the solid represents the limits of possible human understanding, and any bubbles or voids within the solid represent explanatory posits: entities whose claim to reality lies in the fact that the assumption of their existence helps to round out our understanding of the world. The central goal of metaphysics, as I see it, is to make the solid as dense and as spherical as possible: to minimize the ratio of the mysterious to the understood, while also minimizing reliance on explanatory posits.

How does radical phenomenalism compare to other metaphysical theories, relative to this goal?

The main alternatives to phenomenalism are physicalism and dualism. Phenomenalists say that the physical facts supervene on the phenomenological facts, but not vice versa; physicalists say that the phenomenological facts supervene on the physical facts, but not vice versa; dualists say that neither class of facts supervenes on the other.[9]

[7] Among the more counter-intuitive features of relativistic spacetime are the (related) facts that in it, (1) the length of one side of a triangle is not generally less than the combined length of the remaining two sides; (2) the distance between a point *A* and a distinct point *B* can be zero, and the distance between *B* and a distinct point *C* zero, even though the distance between *A* and *C* is something positive; and, (3) every two points are connected by a path with zero length. The situation with quantum mechanics is, if anything, even more extreme. Here, there is not even any consensus on the correct physical interpretation of the basic mathematics of the theory, apart from a general agreement that the correct interpretation, whatever it may be, is unlikely to resonate with ordinary ways of thinking (whence Richard Feynman's quip that "nobody understands quantum mechanics").

[8] The image is Chalmers's: Chalmers (2010b), 381–2. Recently, some writers have tried to parlay the abstract character of the modern scientific image into a form of panpsychism or Berkeleyan idealism: see Strawson (2006), Adams (2007), and Foster (2008); see also Hartshorne (1946). Others conclude that a scientific understanding of the physical world actually compels us to deny the existence of the homely things that make the naive picture so appealing; see, for example, Merricks (2001).

[9] There is, actually, a fourth option worth mentioning: protophenomenalism; see Chalmers (1996a), 153–5, 298–9. A protophenomenalist conjectures that there is a level of reality that serves as a kind of substructure for any traditional dualist, physicalist, or idealist metaphysic. We can understand protophenomenalism as being consistent with dualism: a dualist protophenomenalist would be someone who denied any reduction of physical to phenomenal or phenomenal to physical, but held that phenomenal (and possibly physical) facts reduced to more basic protophenomenal facts. In the same way, protophenomenalism is consistent with the physicalist claim that we can analyze mental facts into physical facts; it's just that if protophenomenalism is true, we can also analyze them into non-physical, protophenomenal facts. And protophenomenalism is consistent with the idealist

Each of these theories has its strengths and weaknesses.

The main strength of dualism is its appeal to common sense. Human beings have a natural inclination to think that conscious experience is something that arises from physical processes (such as brain processes), without *being* a physical process, or any other purely physical kind of thing. Most human beings are dualists by default. That is not an argument for dualism, but it does make dualism an attractive position.

In fact, all of the most powerful arguments advanced in favor of dualism are really only arguments *against* physicalism. They are arguments (like the zombie argument and the knowledge argument) that conclude that the phenomenological facts do not supervene on the physical facts. This conclusion is compatible with idealism; in fact, it goes halfway toward establishing idealism, which is simply the conjunction of this conclusion with the claim that the physical *does* supervene on the phenomenological.[10]

The main weakness of dualism is that it does not do a very good job of minimizing the ratio of the mysterious to the understood. Dualism leaves *two* classes of phenomena fundamentally mysterious: the physical and the phenomenological. Ultimate physical mysteries are, for the dualist, ultimate mysteries *tout court*, as are ultimate phenomenological mysteries. No matter how far the dualist succeeds in identifying lawlike correlations between the mental and the physical, at the end of the day, his account leaves consciousness an enigmatic sideshow of an equally enigmatic main event. Relative to the central goal of metaphysics, dualism puts in a disappointing performance.

What about physicalism? If there is a dominant metaphysical orientation today, this is it. At its most ambitious, physicalism is equivalent to what David Lewis calls "Humean supervenience":

> Humean supervenience . . . is the doctrine that all there is to the world is a vast mosaic of local matters of particular fact, just one little thing and then another . . . We have geometry: a system of external relations of spatiotemporal distances between points . . . And at those points we have local qualities: perfectly natural intrinsic properties which need nothing

claim that the physical facts reduce to phenomenological facts; it's just that these phenomenological facts reduce to further protophenomenal facts, according to a protophenomenalist.

Of course, it is difficult to conceive of non-abstract entities that are neither mental nor physical, and only the strongest arguments could convince us that such entities existed. The fact that there are no such arguments explains why there is no such conviction among philosophers today.

In any case, the effect of introducing a protophenomenal level of reality is not so much to undermine any traditional metaphysical theory, as to give all such theories the status of "surface metaphysics" concerned with the relationship between two classes of entities (mental and physical), at least one of which reduces to something more basic than either.

[10] For the zombie argument, see especially Campbell (1970), 100–4, Kirk (1974), and Chalmers (1996a), 94–9. For the knowledge argument, see especially Jackson (1982).

bigger than a point at which to be instantiated. For short: we have an arrangement of qualities. And that is all. There is no difference without difference in the arrangement of qualities. All else supervenes on that.[11]

Physicalism is sometimes lauded for being a "naturalistic" theory. But what is naturalism? If the hallmark of naturalism is "the supervenience of all upon the microphysical," then radical phenomenalism is clearly not naturalistic. But there is another sense in which radical phenomenalism is at least as naturalistic as the metaphysics of Lewis and his followers. This is the sense in which a naturalist theory shows "great respect for the findings of science (suitably interpreted)," and favors, where possible, the supervenience of all *physical* reality upon the microphysical. To the extent that it is possible to be a naturalist without insisting that physics be treated as metaphysically fundamental, a radical phenomenalist can be a naturalist.[12]

Indeed, if there is anything to the central argument of this book, phenomenalism might do a *better* job than physicalism of respecting the findings of science, when these are taken to include those findings that bear on the fundamental structure of the physical world. Strange to say, the most persuasive argument against physicalism might be physics.

Physicalism and phenomenalism are both forms of ontological monism. There is a clear sense in which both physicalists and phenomenalists say that there is just one kind of thing, at the most fundamental level: physical events for the physicalist, phenomenological potentials for the phenomenalist. In terms of ontological parsimony, neither theory outdoes the other.

But each theory does have one great advantage over the other. For, while both physicalists and phenomenalists meet common sense halfway, they approach the meeting point from opposite directions.

The great advantage of physicalism over phenomenalism is that physicalism, unlike phenomenalism, respects the common sense notion that conscious experience has bodily origins, arising from physical processes in various brains. We tend to think of consciousness as a late entrant to the ontological scene, rather than as the primordial stuff of all reality. Physicalism does not ask us to think otherwise.

The great advantage of phenomenalism over physicalism is that phenomenalism, unlike physicalism, does not ask us to believe that conscious experience is a physical phenomenon. We tend to think of a conscious experience as something over and above any purely physical event, process, or state of affairs. Phenomenalism does not ask us to think otherwise.

[11] Lewis (1986b), ix–x. For a recent study in Lewisian physicalism, see Sider (2011).

[12] The quotes in this paragraph are from Daniel Nolan's unpublished but widely-disseminated *Credo of the Canberra Planners*.

Ultimately, the contest between physicalism and phenomenalism hinges on which of these two advantages is greater. In my opinion, phenomenalism emerges as the clear winner. The idea that there is a deep connection between the world of physics and the world of experience is bound to occur to anyone who makes a sustained effort to unpack the meaning of such everyday statements as "There is a tree over there" or "The table was covered with hors d'oeuvres." For how are we to conceive of a tree or a table whose existence has no potential experiential ramifications? What could it possibly mean to speak of an in-principle imperceptible hors d'oeuvre? Phenomenalism may push the basic idea farther than most people are willing to go, but the idea is at least tolerably clear.

This is in stark contrast to the basic idea behind physicalism. The suggestion that facts about conscious experience are automatically given with purely physical facts (about brain chemistry or whatnot) boggles the mind in a way that phenomenalism does not. There is something almost mystical about it, like being told that a wafer of bread is actually a piece of someone's body. Maybe the suggestion is true, but if so, it is a mind-boggling truth.

Phenomenalism provokes. Physicalism bewilders. Reflecting this difference is the fact that resistance to phenomenalism usually comes as a welter of small- to medium-sized objections (How can phenomenalism account for unperceived things? How can it distinguish reality from a dream? Don't our experiences have to come from *somewhere*?), whereas resistance to physicalism usually takes the form of a single, inexorable objection (If physicalism is true, why is it so easy to conceive of people who are physically just like us, but have no conscious experience?).

It would be absurd for me to claim to have established beyond a reasonable doubt that phenomenalism is superior to all other metaphysical systems. Some small-scale objections are fatal, and some large-scale objections turn out to be surmountable. What I do hope to have accomplished here is the reopening of a neglected third front on the mind–body problem. The fact is that the phenomenalist position was abandoned before it was even completed, much less manned by partisans who could exploit its natural advantages. The old Millian guns have stood silent too long. It is time we heard from them again.

Frank Ramsey sought to capture the spirit of phenomenalism when he wrote:

> My picture of the world is drawn in perspective, and not like a model to scale. The foreground is occupied by human beings, and the stars are all as small as threepenny bits. I don't really believe in astronomy except as a complicated description of human and possibly animal sensation.[13]

[13] Ramsey (1925), 291.

The part about astronomy being a complicated description of sensation is unmistakably phenomenalistic. But there is also a strong hint of Berkeleyanism in Ramsey's remarks—the part about the world being drawn in perspective and the stars being as small as threepenny bits—and this is entirely foreign to phenomenalism. If a star is as small as a threepenny bit because the experience of it takes up so little of my visual field, then what am I to say of an atom, which takes up none of it? But the fact is that although the distant stars correspond to relatively inconspicuous patterns in human experience, what the stars actually *are*, by phenomenalist lights, are *potentials* for experience which do not depend on any human experience for their reality. Here we can do no better than to quote Mill:

> The conception I form of the world existing at any moment, comprises, along with the sensations I am feeling, a countless variety of possibilities of sensation: namely, the whole of those which past observation tells me that I could, under any supposable circumstances, experience at this moment, together with an indefinite and illimitable multitude of others which though I do not know that I could, yet it is possible that I might, experience in circumstances not known to me. These various possibilities are the important thing to me in the world.[14]

Mill describes sensation temporally, but we know how to translate his thought into atemporal terms. The point is that Mill's picture of the world, and the picture that radical phenomenalism presents, is drawn like a model to scale, not in perspective. Or, for a metaphor not available to Mill: the world is an all-encompassing phenomenological hypertext. For the phenomenalist, it is experience all the way down, and it's a long way down.

Earlier, I said that we might think of the intelligibility of the world as a gift to us from the Universe. A religious man might have said: from God. Phenomenalism does not invoke the idea of God, but it would not be too incongruous to think of God as a personification of the phenomenological limit. For this limit plays a role in my theory that certain traditional idealist metaphysics reserve for God: it is the essence of objectivity. For Berkeley, to be objectively real is to be perceived by God; for me, it is to be perceived in the phenomenological limit.

Some say that the greatest mystery about consciousness is the fact that it exists at all. Many have thought that the greatest mystery of all is that anything exists. From the point of view reached in this book, these come to much the same mystery.

[14] Mill (1865/1979), 179–80.

Appendix: objections to relativity

In Section 2.2, I said that the considerations we offer and accept as evidence that one event happened at the same time as another, separate, event were our best guide to the ordinary concept of distal simultaneity. Not everyone agrees with this statement. According to those who do not, considerations of signal isotropy (whether optical, ballistic, or dromedary) pertain at most only to how we figure out whether spatially separated events are simultaneous, and not to what it is for them to *be* simultaneous.

What it is for spatially separated events to *be* simultaneous, in their view, is another matter. It is, for example, for the events both to occur in the presence of a single infinitely fast object—or at least to be such that they could both occur in the presence of such an object, if there were one.[1] Or maybe it is for the two events to be such that they get apprehended in a single act of awareness performed by a being whose powers of perception extend instantaneously throughout the entire cosmos—or at least, to be such that they would get apprehended in a single such act, if there were such a being.[2]

Let us suppose, for the sake of argument, that there *is* a series of signals, each of which propagates instantaneously across the entire cosmos. Still, it must be admitted that we never observe such signals. So even if absolute time exists, we cannot set our clocks to it.

Does this bear only on the verifiability of judgments of absolute distal simultaneity, and not on their truth? But if we can never know whether a judgment of absolute distal simultaneity is true, then facts of absolute distal simultaneity, if there are such facts, are of no practical interest to us. If you tell an ordinary person that there is no way to know that two events that happened in different cities happened at the same time, he will just become confused, or else take you to be peddling some bogus hyperbolic doubt. Yet, there is not, in fact, any way to know that events that happen in different cities happen *absolutely* simultaneously. This strongly suggests that absolute distal simultaneity is not part of our common sense conception of time.

The basic point goes back to Poincaré, who saw that duration, succession, and other temporal concepts had to be of such a nature to permit knowledge of their successful application to the world, since we actually do apply them successfully, and know it. Poincaré recognized that only a defective account of time could entail that we never knew that two

[1] This is the image of absolute simultaneity we get from Lorentz, when he confesses that he "finds a certain satisfaction in the older interpretations, according to which . . . space and time can be sharply separated, and simultaneity spoken of without further specification. In regard to this last point, one may perhaps appeal to our ability to imagine arbitrarily high speeds. Therewith one comes very close to the concept of absolute simultaneity." Lorentz (1914), 23.

[2] This is the image of absolute simultaneity we get from Poincaré, when he discusses (in a critical vein) the idea of "a sort of great consciousness which should see all, and which should classify all *in its time*, as we classify, *in our time*, the little we see." Poincaré (1913), 229. For a discussion that takes this image seriously, see Craig (2001), 171–94.

separate events occurred simultaneously, just as only a defective account of knowledge could entail that we never knew anything at all.[3]

Acknowledging that our ordinary concept of time gets revealed in our practices of making and evaluating temporal judgments does not commit us to verificationism—the doctrine that a statement is meaningful only to the extent that we can conceive of proving ("verifying") or disproving it. The doctrine is obscure—what constitutes proof? And don't I already have to know what a statement means, before I can consider whether any proof or disproof of it is possible?—but we do not need to embrace it in order to acknowledge that part of what makes our ordinary concept of simultaneity our *ordinary* concept is that we can recognize correct applications of it as such. This is not a verificationist idea—or if it is, that just shows that not all verificationist ideas are bad.

In my view, the relativistic conception of simultaneity is a natural extrapolation from our ordinary, pre-theoretical concept of simultaneity. The key insight was Galileo's, that the simultaneity of two events comes to the possibility of measuring them as occurring at the same time using synchronized clocks. To say that two clocks are "synchronized" is to say that they tell the same times at the same times, and the meaning of this statement is contained in the methods by which we actually synchronize clocks, or are prepared to synchronize them. If there are spatially separated events that are simultaneous in some other, "absolute" sense unconnected with these methods, then that is a sense that is quite alien to our ordinary understanding of simultaneity, according to which we very often correctly judge spatially separated events to have occurred at the same time.

Some authors emphasize that the observations that lead us to a relativistic conception of spacetime are logically compatible with an absolute conception. The earliest and clearest example of this approach is due to George FitzGerald and Hendrik Lorentz, who pointed out that the observations that support a relativistic conception of simultaneity equally support a physics according to which simultaneity is absolute, but objects and processes change their spatial and temporal magnitudes depending on their states of motion relative to absolute space.[4]

It is true that the observational data do not strictly compel us to adopt a relativistic conception of spacetime. It is also true that the observational data do not strictly compel us to adopt a heliocentric conception of the Solar System. The question is whether we have any better reason to doubt the relativistic conception than to doubt heliocentrism.

Today, the most common reason for doubting the relativistic conception is that we need to posit some kind of anti-relativistic superluminal communication, in order to account for certain surprising statistical correlations—the so-called Einstein-Podolsky-Rosen (or "EPR") correlations. These are best explained with an analogy.[5]

Suppose that you work in a laboratory in London, while I work in a similar laboratory in Seattle. Each of us has a big crate of bottles, sent to us from a manufacturer in Atlanta. The

[3] For Poincaré's observations, see Poincaré (1913). James Cargile makes the point as regards knowledge in general in Cargile (1998), 252.

[4] See FitzGerald (1889), Lorentz (1937), and Lorentz (1916), 230.

[5] The analogy I am about to give follows the pattern of Mermin (1985) and Brown (2010), 650–6. In their original discussion, Einstein, Podolsky, and Rosen present the EPR correlations as a challenge to quantum theory; see Einstein et al. (1935). In Bell (1964), John Bell presents the correlations as a challenge to relativistic mechanics.

bottles in my crate are numbered from one up to some large number (equal to the number of bottles in my crate), and each bottle contains a single pill. Your crate contains the same number of bottles, also numbered consecutively, and also containing one pill each.

Both labs have a large supply of rats, canaries, and goldfish. Each of us has been instructed in advance to open the numbered bottles one by one, feeding each bottle's pill to an animal of our choosing. It is up to me whether I feed any given pill to a rat, a canary, or a fish. If I want, I can feed all my pills to rats, or half to rats and half to fish, or follow any other policy I like, including dispensing pills to animals at random. You are likewise free to choose which kind of animal gets the pill in each of your numbered bottles. However, at no point do we communicate with one another; in particular, we make no attempt to coordinate our choices concerning which pills we give to which animals.

Now, it turns out that the factory in Atlanta produces a variety of pills: a placebo which has no effect on any animal, a poison that is lethal to all animals, and a range of other pills, each of which is lethal to some, but not all, species. Let "R," "B," and "F" designate harmlessness to rats, birds, and fish, respectively, and "$\overline{R}$," "$\overline{B}$," and "$\overline{F}$" designate lethality to these respective species. So, for example, "RBF" designates a placebo, "$\overline{RBF}$" a general biocide, and "$R\overline{BF}$" a pill that is harmless to rats, but lethal to both birds and fish.

The Atlanta factory produces eight kinds of pills (unmarked and outwardly indistinguishable from one another) characterized by the following pharmaceutical properties:

$$RBF$$
$$RB\overline{F}$$
$$R\overline{B}F$$
$$\overline{R}BF$$
$$R\overline{BF}$$
$$\overline{RB}F$$
$$\overline{R}B\overline{F}$$
$$\overline{RBF}$$

At a pre-arranged time, both of us—you in London, I in Seattle—begin to open our bottles and feed their contents to our lab animals, with the following results.

Half of the rats that I feed pills to die, half of the canaries I feed pills to die, and half of the fish I feed pills to die. Likewise, half of the rats that you feed pills to die, half of the canaries you feed pills to die, and half the fish you feed pills to die.

However, when we compare our results later on, we find a curious correlation. Recall that our bottles were labeled with numbers: I had a bottle labeled "1," you had a bottle labeled "1"; I had a bottle labeled "2," you had a bottle labeled "2"; etc. For each number, *n*, if we both fed the contents of our number *n* bottles to the same kind of animal, exactly one of the animals died. In other words, in no case in which we chose to feed the *n*th bottle's pill to the same kind of animal did we get the same result: if you fed your 17th bottle's pill to a fish, and I fed my 17th bottle's pill to a fish, either your fish died and mine lived or vice versa; if you fed your 1378th bottle's pill to a rat, and I fed my 1378th bottle's pill to a rat, either your rat died and mine lived or vice versa; and so forth. We summarize these results in Table 13.1, where percentages indicate the frequency with which our choices (of what kind of animal to give a pill) got the same result (both animals surviving, or both dying).

Table 13.1 Results of the experiment.

	You feed pill to a rat	You feed pill to a bird	You feed pill to a fish
I feed pill to a rat	0%	75%	75%
I feed pill to a bird	75%	0%	75%
I feed pill to a fish	75%	75%	0%

These results might lead us to suspect that the pills were concocted and bottled at the factory in such a way as to ensure that the percentages came out as they did. Given that we, in our respective labs, are free to administer the pills however we like, and are not in communication with one another, the only way to guarantee these results would be to make sure that each pair of like-numbered bottles is such that one bottle of the pair but not the other contains a pill lethal to rats, one bottle of the pair but not the other contains a pill lethal to birds, and one bottle of the pair but not the other contains a pill lethal to fish. So, for example, if my bottle #44 contains a pill of characteristic $RB\overline{F}$, the pill that your bottle #44 contains must have characteristic $\overline{RB}F$; only thus can the folks in Atlanta ensure that we won't get the same results if we choose to feed the respective contents of our #44 bottles to the same kind of animal.

There are eight ways to fill a given pair of bottles to ensure that we never get the same result if we happen to feed same-numbered pills to the same kind of animal (see Table 13.2). The question we must ask is whether the people in Atlanta could have filled the bottles they sent to us (in London and Seattle) according to some combination of schemes (1) through (8).

Clearly, it would not do for them simply to fill every pair of bottles in accordance with scheme (1). While that would certainly guarantee that our labs would get different results whenever we fed pills from like-numbered bottles to conspecific animals, it would make it impossible for us to get identical results (both animals live, or both die) in cases in which, for example, I feed my *n*th pill to a fish, and you feed your *n*th pill to a rat—whereas in fact we do get the same results in such cases 75 percent of the time (as indicated in the lower left

Table 13.2 Bottling schemes.

	My n^{th} bottle	Your n^{th} bottle
(1)	RBF	$\overline{RBF}$
(2)	$RB\overline{F}$	$\overline{RB}F$
(3)	$R\overline{B}F$	$\overline{R}B\overline{F}$
(4)	$R\overline{BF}$	$\overline{R}BF$
(5)	$\overline{R}BF$	$R\overline{BF}$
(6)	$\overline{RB}F$	$RB\overline{F}$
(7)	$\overline{R}B\overline{F}$	$R\overline{B}F$
(8)	$\overline{RBF}$	RBF

cell of Table 13.1). Nor, for exactly parallel reasons, would it do to fill every pair of bottles according to any other single scheme.

If there is any way for the Atlanta bottlers to guarantee the results shown in Table 13.1, it must be through filling bottles in accordance with some *combination* of schemes (1) through (8).

This, however, is demonstrably impossible.

Let P(1), P(2), P(3), . . . and P(8) be the percentage of bottles filled in accordance with scheme (1), scheme (2), scheme (3), . . . and scheme (8), respectively. Since we know that the Atlanta people have to fill each pair of bottles according to one of these schemes (in order to ensure that the labs never get the same result when feeding like-numbered pills to conspecific animals), we know that:

$$P(1) + P(2) + P(3) + P(4) + P(5) + P(6) + P(7) + P(8) = 100\%$$

Furthermore, we know that:

$$P(3) + P(4) + P(5) + P(7) = 75\%$$

We know this, because any pair of bottles filled according to scheme (3), (4), (5), or (7) is such that if one of us feeds a bird, and the other a rat, from his bottle from that pair, we get the same result (both of us kill an animal, or neither of us kills an animal). Therefore, if more than 75 percent of paired bottles were filled according to schemes (3), (4), (5), or (7), our results would agree in more than 75 percent of the cases represented by the top center and left middle cells of Table 13.1. Likewise, if less than 75 percent of paired bottles were filled according to schemes (3), (4), (5), or (7), our results would agree in less than 75 percent of the cases represented by the aforementioned cells.

By exactly parallel reasoning, we know that:

$$P(2) + P(4) + P(5) + P(6) = 75\%$$

and that:

$$P(2) + P(3) + P(6) + P(7) = 75\%$$

Putting everything together, we have:

$$[P(3) + P(4) + P(5) + P(7)] + [P(2) + P(4) + P(5) + P(6)] + [P(2) + P(3) + P(6) + P(7)] = 75\% + 75\% + 75\%$$

which reduces to:

$$2[P(2) + P(3) + P(4) + P(5) + P(6) + P(7)] = 225\%$$

which gives us, finally:

$$P(2) + P(3) + P(4) + P(5) + P(6) + P(7) = 112.5\%$$

The upshot is that in order for the people in Atlanta to fill the bottles so as to guarantee the results summarized in Table 13.1—in order for them to "rig the system" to ensure that these are the percentages that emerge—they would have to fill 112.5 percent of the bottle pairs in accordance with schemes (2) through (7). Since this is impossible, it is impossible for the people in Atlanta to rig the system by choosing a suitable combination of bottle-filling schemes.

In the foregoing analogy, pairs of pills in like-numbered bottles stand for "prepared electrons"—electrons that have been polarized with a powerful magnet. Feeding a pill to an animal is analogous to subjecting a prepared electron to one of three tests to determine its spin angular momentum along a chosen axis. The percentages given in Table 13.1 are identical to those that actual experiments on electrons yield. Finally, the lack of communication between the two laboratories represents the fact that the actual tests performed on paired electrons are carried out at a spacelike separation from one another, precluding any (non-superluminal) communication between them.

The EPR correlations are just the subatomic analogue of the fictional correlations recorded in Table 13.1. The correlations are surprising, because they subvert the most natural explanation of the fact that physicists conducting spacelike separated experiments never get the same result when they subject a pair of prepared particles to the same tests. The most natural explanation for this is that the process of preparing the particles for subsequent testing somehow pre-programmed the particles to react to subsequent testing in such a way as to give the results that we actually observe. Yet, as in the case of the pills and the animals, straightforward statistical considerations rule out any such explanation.

The EPR correlations do not directly imply any anti-relativistic superluminal signalling, since it is impossible to exploit the correlations to transmit information. Still, there are those who suggest that the existence of the correlations must have some *explanation,* and that the best explanation is bound to invoke some kind of anti-relativistic communication between spacelike separated events.

This is an inference to the best explanation. It is dubious, on two counts.

First, it is not obvious that the EPR correlations require any explanation. We could take the correlations as a fundamental natural regularity, and the (non-classical) laws that describe this regularity as fundamental natural laws. From this point of view, the search for a mechanism to *explain* the correlations is as badly conceived as the ill-fated search for a mechanism to explain gravitational phenomena.[6]

Second, even if we do demand an explanation for the correlations, it is far from clear that the *best* explanation must involve faster-than-light causal influence. This is an area of ongoing investigation, but there are at least three live options worth mentioning that are consistent with relativity.

One is that the test-particles are pre-programmed—not to yield certain measurements upon being subjected to certain tests, but to cause experimenters to apply certain tests to them. If the particles get to choose which tests the experimenters perform on them, then the particles can, as it were, conspire in advance to require their respective testers

[6] For a detailed criticism of the inference to the best explanation, see Fine (1989). The most famous rejecter of the inference is Niels Bohr.

to take certain measurements of them—measurements guaranteed to yield the observed EPR correlations. Physicists sometimes object to this explanation on the grounds that it deprives experimenters of free will, but this is a weak objection, for reasons that are too familiar to rehearse here.[7]

A second possibility is that the manner in which the particles are prepared (how a given pair of bottles gets filled) depends on which tests the experimenters choose to perform (which animals they choose to feed from those bottles). In other words, the correlations might be explained by the fact that the experimenters' choices of which tests to perform caused certain particles rather than others (or particles with certain features rather than others) to be the subjects of those tests. This explanation has the disadvantage of implying backwards causation, whereby later events cause earlier ones. This is a disadvantage not because backwards causation is incoherent—it is not incoherent, or at least not obviously so—but because it implies the existence of causal loops. Whether such loops can be eliminated, or their unwelcome implications mitigated, is a matter of continuing debate.[8]

A third possibility is that the experimenters' choices as well as the events involved in the preparation of the test-particles are common effects of events from which they are both lightlike separated. (See Fig. 13.1, in which *A*, *S*, and *L* are a preparation of test-particles in Atlanta, a testing of prepared particles in Seattle, and a testing of prepared particles in London, respectively.) This has an advantage over ordinary backwards causation theories, in that it requires causation only between spacetime co-located events, which is not "backward" in any absolute sense—indeed, in absolute terms, such causation is perfectly local, there being no spacetime distance between any of the causally connected events. However, as with theories that invoke backwards causation between timelike separated events, this theory would have to be refined to avoid causal loops.[9]

Even if there were a defensible inference to the best explanation for the existence of a non-relativistic conception of spacetime, it is hard to see what further advantages such a conception could possibly have over the relativistic conception. Even if we had reason to think that absolute distal simultaneity existed, we could never have any reason to believe, of any two spatially separated events, that they were, in fact, absolutely simultaneous. As Craig Callender points out, this means that it is impossible to analyze any observable phenomenon in terms of absolute distal simultaneity, since any such analysis would presuppose the possibility of identifying actual instances of absolute distal simultaneity. (This includes introspectible phenomena, such as our own conscious experiences.)[10]

[7] As any decent introductory philosophy textbook explains, the claim that we have free will is controversial, as is the claim that free will is incompatible with the sort of determinism that the proposed explanation of the EPR correlations implies.

[8] For the backwards causation view, see Price (1996), and, for relevant discussion and criticism, Arntzenius (1994), Berkovitz (2002), and Berkovitz (2008).

[9] The proposal glossed here has an affinity with the Wheeler-Feynman "absorber" theory of radiation, as well as with the position developed by Tetrode in the paper which is the source of the epigraph of Chapter 2: see especially Tetrode (1922), 324–8. For an illuminating discussion, see Brown (2010), 673–99.

[10] See Callender (2008b). The basic point, of course, goes back to Poincaré.

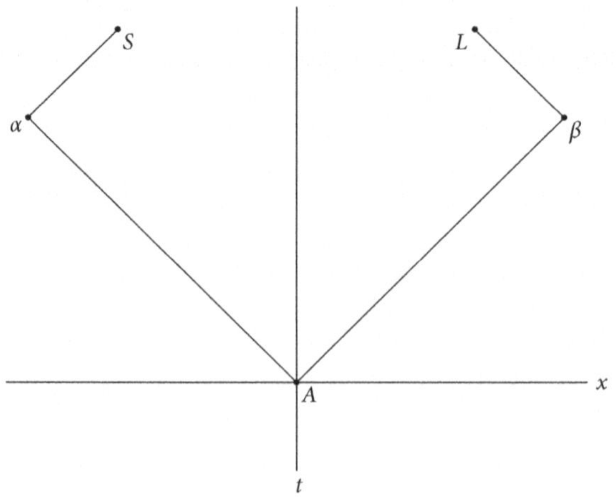

Figure 13.1 S an effect of α. L an effect of β. A the combined effect of α and β.

Although the most common challenges to relativity are based on the EPR correlations, one might attempt a similar challenge by reference to conscious experience. One might argue that an audiovisual experience, like an experience as of a simultaneous flash and bang, consists of the simultaneous occurrence of a visual experience (*Flash!*) in one part of your brain, and an auditory experience (*Bang!*) in a different part of your brain. Since the existence of the audiovisual experience is an absolute fact, so too, one might reason, must be the relation of simultaneity between the auditory and visual experiences. Since there is only one frame of reference in terms of which of the two experiences are simultaneous, one might conclude that this frame of reference—call it F_P—is "preferred," in the sense that a complete and accurate description of the world (which must include a description of my mental life) is possible only in terms of F_P.

Now consider your counterpart on Twin Earth, and suppose that Twin Earth is in motion in terms of the frame of reference in terms of which Earth is at rest. In terms of F_P, your Twin's *Flash!* and *Bang!* experiences occur in temporal succession, separated by an interval of, let us say, five years. Thus, if F_P alone permits a complete and accurate description of reality, it must be that your Twin's experiences are, as a matter of absolute fact, separated by a five-year interval. In that case, your Twin does not have an experience as of simultaneous flash and bang, but only at most an experience as of a flash followed by a bang.[11]

Yet, in terms of a frame of reference that represents Twin Earth as stationary, things on Twin Earth are exactly the same as we describe things to be here on Earth (in terms of a frame of reference that represents Earth as stationary). There is your Twin, over on Twin

[11] Could one say that a *Flash!* experience and a *Bang!* experience occurring (absolutely) five years apart could constitute an experience as of a simultaneous flash and bang? Suppose one does say that. Then one has no good reason to think that an experience as of a simultaneous flash and bang consists of absolutely simultaneous *Flash!* and *Bang!* experiences, as this whole argument for a preferred frame of reference assumes.

Earth, with a brain just like yours, in which the same physical processes occur as in your brain, and out of whose mouth come words indistinguishable from those that you utter. And yet—according to the theory that your audiovisual experience involves the occurrence of separate but absolutely simultaneous phenomenal events—your Twin does not have any experience as of simultaneous flash and bang.

More generally, positing F_P would require us to say that the phenomenal features of your conscious life differed dramatically from those of your Twin, and that if you and your Twin were to switch places (your Twin emigrating to Earth, and you to Twin Earth), each of you would have dramatically different experiences from what he had before, despite having the same body embedded in an environment qualitatively indistinguishable from his old environment. This phenomenological geocentrism has even less to recommend it than its more familiar (and thoroughly discredited) astronomical cousin.[12]

Not only do contemporary opponents of relativity tend to overestimate the benefits of abandoning the relativistic point of view in favor of a non-relativistic theory: they also underestimate the costs of abandoning relativity. When Lorentz proposed to retain the absolute conception of time by supposing that objects change their lengths (and processes their durations) depending on their velocities with respect to absolute space, he was, in effect, proposing to reject the principle of inertia—he was suggesting that it is *false* that inertia is homogenous and isotropic.

This fact is easy to overlook, since it is easy to forget that the second "postulate" of special relativity—that in terms of every inertial frame of reference, light travels at the same speed regardless of its direction and regardless of the state of motion of its source—is actually an empirically established truth. It is a truth established by the consideration that the onboard clocks of N(−1,0,0) and N(1,0,0) [and N(−2,0,0) and N(2,0,0), etc.] give the same reading when light from station N(0,0,0) reaches them, these stations having been positioned and their clocks synchronized on the assumption that mechanical inertia is homogenous and isotropic (and similarly for the stations in the Perseid array). This is not to say that a belief in frame-independent lightspeed isotropy is forced upon us by observation, only that it is forced upon us by observation *together with* our decision to regard mechanical inertia as homogenous and isotropic.

These considerations also have a bearing on the mistaken idea that empirical observation establishes only that the round-trip speed of light is constant in terms of every inertial coordinate system, the constancy of the one-way speed being no more than an optional simplifying assumption. The truth is that the invariance of the one-way speed is as firmly established by observation as the invariance of the round-trip speed; this is again illustrated by the example of *Nike* and *Perseus*. If we want to identify something as a simplifying assumption here, it is the assumption that inertially synchronized clocks really are synchronized; that is, in effect, the assumption that the principle of inertia is true.[13]

12 Here, "switching places" is shorthand for switching statuses as regards state of motion in terms of F_P. For additional criticism of the idea that phenomenology provides a basis for identifying a preferred frame of reference, see Lee (2007), 354–8.

13 For the suggestion that the constancy of the one-way speed of light is a mere simplifying assumption (and a false assumption at that), see Tooley (1997), 341–8.

If the principle of inertia is true, then the speed of every photon at any time or place is *c*. That's because the speed of every photon at any time or place is *c* in terms of every inertial frame of reference, and an inertial frame of reference is simply a frame of reference in which the principle of inertia is true. So if the speed of light is not a constant—if the speed of some photon at some time and place is greater or less than *c*—then the principle of inertia is false. If we say, with Lorentz and FitzGerald, that photons traveling through empty space travel more slowly in one direction than another, then we must reject the principle of inertia. Neo-Lorentzian mechanics is not a return to Newton: it is a return to Aristotle.

Considering how fundamental the principle of inertia is to our scientific understanding of the world, rejecting the principle should be a measure of very last resort. And given that we cannot abandon the relativistic conception of spacetime without abandoning the principle of inertia, we should be wary of any attempt to cast doubt on the relativistic conception for the purpose of shoring up one or another philosophical position.[14]

[14] In general relativity, inertia plays a different role from the one it plays in Newtonian mechanics and special relativity, but it still plays a role. Instead of describing the world in terms of an inertial coordinate system, general relativity describes it in terms of a more complex spatiotemporal continuum partly characterized by the fact that in smaller regions of the continuum, inertia is more nearly homogenous and isotropic than in larger regions of it, and, in the infinitesimal limit, perfectly homogenous and isotropic.

Bibliography

Adams, Robert M. (2007) "Idealism vindicated." In: van Inwagen, Peter, and Zimmerman, Dean (eds), *Persons: Human and Divine*. Oxford: Oxford University Press, pp. 35–54.

Armstrong, David (1961) *Perception and the Physical World*. London: Routledge & Kegan Paul.

Arntzenius, Frank (1994) "Spacelike connections." *British Journal for the Philosophy of Science* 45, 1: 201–17.

Ayer, A.J. (1946–1947) "Phenomenalism." *Proceedings of the Aristotelian Society* 47: 163–96.

Barbour, Julian (1999) *The End of Time: The Next Revolution in Physics*. Oxford: Oxford University Press.

Bayne, Tim (2010) *The Unity of Consciousness*. Oxford: Oxford University Press.

Bayne, Tim, and Chalmers, David (2010) "What is the unity of consciousness?" In: Chalmers, David, *The Character of Consciousness*. Oxford: Oxford University Press, pp. 497–539.

Bell, John S. (1964) "On the Einstein Podolsky Rosen Paradox." *Physics* 1, 3: 195–200.

Berkeley, George (1710/1982) *A Treatise Concerning the Principles of Human Knowledge*, Winkler, K. (ed.). Indianapolis, IN: Hackett Publishing.

Berkovitz, Joseph (2002) "On causal loops in the quantum realm." In: Placek, Tomasz, and Butterfield, Jeremy (eds), *Non-locality and Modality*. Dordrecht: Kluwer, pp. 235–57.

Berkovitz, Joseph (2008) "On predictions in retro-causal interpretations of quantum mechanics." *Studies in History and Philosophy of Modern Physics* 39: 709–35.

Block, Ned (2007) "Troubles with functionalism." In: *Consciousness, Function, and Representation: Collected Papers*, Volume 1. Cambridge, MA: MIT Press, pp. 63–101.

Boscovich, Roger Joseph (1763/1922) *A Theory of Natural Philosophy*, Child, J.M. (trans.). Chicago and London: Open Court Publishing Company.

Brentano, Franz (1874/1995) *Psychology from an Empirical Standpoint*. London: Routledge.

Broad, C.D. (1923) *Scientific Thought*. London: Kegan Paul, Trench, Trubner & Co.

Brown, Kevin (2010) *Reflections on Relativity*. Copyright: Kevin Brown.

Burge, Tyler (1977) "Individualism and the mental." *Midwest Studies in Philosophy* IV: 73–121.

Callender, Craig (2004) "There is no puzzle about the low-entropy past." In: Hitchcock, Christopher (ed.), *Contemporary Debates in Philosophy of Science*. Malden: Blackwell, pp. 240–55.

Callender, Craig (2008a) "The common now." *Philosophical Issues* 18: 339–61.

Callender, Craig (2008b) "Finding 'real' time in quantum mechanics." In: Craig, William Lane, and Smith, Quentin (eds), *Einstein, Relativity, and Absolute Simultaneity*. New York: Routledge, pp. 50–72.

Callender, Craig (2010) "Is time an illusion?" *Scientific American*, June: 59–65.

Campbell, Keith (1970) *Body and Mind*. London and Toronto: Macmillan.

Campbell, Scott (2001) "Persons and substances." *Philosophical Studies* 3: 253–67.

Cargile, James (1997) "On the burden of proof." *Philosophy* 72, 279: 59–83.

Cargile, James (1998) "The problem of induction." *Philosophy* 73, 284: 247–75.

Carnap, Rudolf (1963) "Rudolf Carnap: Intellectual Autobiography." In: Schilpp, Paul Arthur (ed.), *The Philosophy of Rudolf Carnap*. LaSalle, IL: Open Court, pp. 1–83.

Carnap, Rudolf (1967) *The Logical Structure of the World and Pseudoproblems in Philosophy*. Berkeley and Los Angeles, CA: University of California Press.

Casati, Roberto, and Varzi, Achille C. (2001) "That useless time machine." *Philosophy* 76, 298: 581–3.

Chalmers, David (1996a) *The Conscious Mind: In Search of a Fundamental Theory*. Oxford: Oxford University Press.

Chalmers, David (1996b) "Does a rock implement every finite-state automaton?" *Synthese* 108, 3: 309–33.

Chalmers, David (2010a) "The Matrix as metaphysics." In: *The Character of Consciousness*. Oxford: Oxford University Press, pp. 455–94.

Chalmers, David (2010b) "Perception and the fall from Eden." In: *The Character of Consciousness*. Oxford: Oxford University Press, pp. 381–454.

Chalmers, David (2012) *Constructing the World*. Oxford: Oxford University Press.

Chalmers, David "The combination problem for panpsychism." <http://consc.net/papers/combination.pdf> [accessed June 12, 2014].

Chisholm, Roderick (1976) *Person and Object: A Metaphysical Study*. LaSalle, IL: Open Court Publishing.

Christenson, J.H., Cronin, J.W., Fitch, V.L., and Turlay, R. (1964) "Evidence for the 2π Decay of the K_2^0 Meson." *Physical Review Letters* 13, 4: 138–40.

Churchland, Paul (1985) "The ontological status of observables: in praise of the superempirical virtues." In: Churchland, Paul, and Hooker, Clifford (eds), *Images of Science: Essays on Realism and Empiricism*. Chicago, IL: University of Chicago Press, pp. 35–47.

Collins, John M. "Epistemic closure principles." <http://www.iep.utm.edu/epis-clo/> [accessed June 9, 2014].

Craig, William Lane (2001) *Time and the Metaphysics of Relativity*. Dordrecht: Kluwer Academic Publishers.

Crane, Tim (1991) "All the difference in the world." *Philosophical Quarterly* 41, 162: 1–25.

Crowther, Thomas (2009) "Watching, sight, and the temporal shape of perceptual activity." *The Philosophical Review* 118, 1: 1–27.

Dainton, Barry (2006) *Stream of Consciousness: Unity and Continuity in Conscious Experience*. New York: Routledge.

Dainton, Barry (2008) *The Phenomenal Self*. Oxford: Oxford University Press.

Dainton, Barry, and Bayne, Tim (2005) "Consciousness as a guide to personal persistence." *Australasian Journal of Philosophy* 85, 4: 549–71.

DeRose, Keith (1995) "Solving the Skeptical Problem." *Philosophical Review* 104, 1: 1–52.

Descartes, René (1641/1984) "Meditations on First Philosophy." In: Cottingham, J., et al. (eds), *The Philosophical Writings of Descartes*, Volume II. Cambridge: Cambridge University Press, pp. 3–62.

Dretske, Fred (1981) *Knowledge and the Flow of Information*. Cambridge, MA: MIT Press.

Dretske, Fred (2003) "Experience as representation." *Philosophical Issues* 13, 1: 67–82.

Dretske, Fred (2005) "The case against closure." In: Steup, Matthias, and Sosa, Ernest (eds), *Contemporary Debates in Epistemology*. Malden: Blackwell Publishing, pp. 13–26.

Dummett, Michael (1978) "Realism." In: *Truth and Other Enigmas*. Cambridge, MA: Harvard University Press, pp. 145–65.

Earman, John (1995) *Bangs, Crunches, Whimpers, and Shrieks: Singularities and Acausalities in Relativistic Spacetimes*. Oxford: Oxford University Press.

Edelglass, William, and Garfield, Jay L. (2009) *Buddhist Philosophy: Essential Readings*. New York: Oxford University Press.

Einstein, Albert (1905/1952) "On the electrodynamics of moving bodies." In: *The Principle of Relativity*. New York: Dover Publications, pp. 35–65.

Einstein, Albert, Podolsky, Boris, and Rosen, Nathan (1935) "Can quantum-mechanical description of physical reality be considered complete?" *Physical Review* 47, 10: 777–80.

Farkas, Katalin (2008) *The Subject's Point of View*. Oxford: Oxford University Press.

Field, Hartry (1978) "Mental representation." *Erkenntnis* 13, 1: 9–61.

Fine, Arthur (1989) "Do correlations need to be explained?" In: Cushing, James T., and McMullin, Ernan (eds), *Philosophical Consequences of Quantum Theory*. Notre Dame, IN: University of Notre Dame Press, pp. 175–94.

FitzGerald, George Francis (1889) "The ether and the Earth's atmosphere." *Science* 13, 328: 390.

Fodor, Jerry (1987) *Psychosemantics*. Cambridge, MA: MIT Press.

Foster, John (1979) "In self-defence." In: Macdonald, G.F. (ed.), *Perception and Identity*. London: Macmillan, pp. 161–85.

Foster, John (1982) *The Case for Idealism*. London: Routledge and Kegan Paul.

Foster, John (1991) *The Immaterial Self*. London: Routledge.

Foster, John (2008) *A World for Us: The Case for Phenomenalistic Idealism*. Oxford: Oxford University Press.

van Fraassen, Bas (1985) "Empiricism in the philosophy of science." In: Churchland, Paul, and Hooker, Clifford (eds), *Images of Science: Essays on Realism and Empiricism*. Chicago, IL: University of Chicago Press, pp. 245–308.

Futch, Michael J. (2008) *Leibniz's Metaphysics of Time and Space*. Boston Studies in the Philosophy of Science, Volume 258. Dordrecht, Boston: Springer.

Galileo, Galilei (1638/1954) *Dialogues Concerning Two New Sciences*, Crew, Henry, and de Salvio, Alfonso (trans.). New York: Dover Publications.

Galison, Peter (2003) *Einstein's Clocks and Poincaré's Maps: Empires of Time*. New York: W.W. Norton & Company.

Garfield, Jay (1995) *The Fundamental Wisdom of the Middle Way: Nagarjuna's Mulamadhyamakakarika*. New York: Oxford University Press.

Garfield, Jay (forthcoming) *Engaging Buddhism: Why It Matters to Philosophy*. New York: Oxford University Press.

Gauss, Carl Friedrich (1900) *Werke: Achter Band*. Göttingen: Königlichen Gesellschaft der Wissenshaften.

Gauss, Carl Friedrich (1902) *General Investigations of Curved Surfaces of 1827 and 1825*, Caddall Morehead, Ames, and Miller Hiltebeitel, Adam (trans.) Princeton, NJ: Princeton University Library.

Geroch, Robert (1978) *General Relativity from A to B*. Chicago, IL: University of Chicago Press.

Gödel, Kurt (1949) "An example of a new type of cosmological solution of Einstein's field equations of gravitation." *Reviews of Modern Physics* 21, 3: 447–50.

Gould, Rupert T. (1923) *The Marine Chronometer: Its History and Developments*. London: J.D. Potter.

Graff, Delia (2001) "Phenomenal continua and the sorites." *Mind* 110, 440: 905–35.

Gustafsson, Johan E. (2011) "Phenomenal continuity and the bridge problem." *Philosophia* 39, 2: 289–96.

Hacking, Ian (1985) "Do we see through a microscope?" In: Churchland, Paul, and Hooker, Clifford (eds), *Images of Science: Essays on Realism and Empiricism*. Chicago, IL: University of Chicago Press, pp. 132–52.

Hájek, Alan. *Most Counterfactuals are False*. <http://philrsss.anu.edu.au/people-defaults/-alanh/papers/MCF.pdf> [accessed June 12, 2014].

Hartshorne, Charles (1946) "Leibniz's greatest discovery." *Journal of the History of Ideas* 7, 4: 411–21.

Hawrylycz, Michael J., and Lein, Ed S., et al. (2012) "An anatomically comprehensive atlas of the adult human brain transcriptome." *Nature* 498, 7416: 391–9.

Hawthorne, John (2005) "The case for closure." In: Steup, Matthias, and Sosa, Ernest (eds), *Contemporary Debates in Epistemology*, Malden: Blackwell Publishing, pp. 26–43.

Healey, Richard (2007) *Gauging What's Real: The Conceptual Foundations of Contemporary Gauge Theories*. Oxford: Oxford University Press.

Hoerl, Christoph (2009) "Time and tense in perceptual experience." *Philosopher's Imprint* 9, 12: <http://quod.lib.umich.edu/p/phimp/3521354.0009.012/1/–time-and-tense-in-perceptual-experience?rgn = full + text;view = image;q1 = Time + and + tense+in+perceptual+experience> [accessed September 25, 2014].

Horgan, Terence, and Tienson, John (2002) "The intentionality of phenomenology and the phenomenology of intentionality." In: Chalmers, David J. (ed.), *Philosophy of Mind: Classical and Contemporary Readings*. New York: Oxford University Press, 520–33.

Hume, David (1739/1978) *A Treatise of Human Nature*. Oxford: Clarendon Press.

Hume, David (1779/2007) *Dialogues Concerning Natural Religion*. Cambridge: Cambridge University Press.

Hurley, Susan L. (1998) *Consciousness in Action*. Cambridge, MA: Harvard University Press.

van Inwagen, Peter (2009) *Metaphysics*, 3rd edn. Boulder, CO: Westview Press.

Jackson, Frank (1973) "Is there a good argument against the incorrigibility thesis?" *Australasian Journal of Philosophy* 51, 1: 51–62.

Jackson, Frank (1982) "Epiphenomenal qualia." *The Philosophical Quarterly* 32, 127: 23–36.

Jackson, Frank (1998) *From Metaphysics to Ethics: A Defense of Conceptual Analysis*. Oxford: Clarendon Press.

Kant, Immanuel (1770/1894) *Kant's Inaugural Dissertation of 1770*, Eckoff, Walter J. (ed.). New York: Columbia College.

Kant, Immanuel (1781/1998) *Critique of Pure Reason*, Guyer, P., and Wood, A.W. (eds). Cambridge: Cambridge University Press.

Kant, Immanuel (1955) *Kant's gesammelte Schriften*, XXIII. 3. Berlin: Walter de Gruyter.

Kirk, Robert (1974) "Sentience and behavior." *Mind* 83, 329: 43–60.

Kriegel, Uriah (2003) "Is intentionality dependent upon consciousness?" *Philosophical Studies* 116, 3: 271–307.

Kriegel, Uriah (2013) "A hesitant defense of introspection." *Philosophical Studies* 165, 3: 1165–76.

Langton, Rae (1998) *Kantian Humility: Our Ignorance of Things in Themselves*. Oxford: Clarendon Press.

Lee, Geoffrey (2007) "Consciousness in a space-time world." *Philosophical Perspectives* 21: 341–74.

Lee, Tsung-Dao, and Ning Yang, Chen (1956) "Question of parity conservation in weak interactions." *Physical Review* 104, 1: 254–8.

Leibniz, G.W. (1680/1999) "De Notionibus Omnia Quae Cogitamus Contentibus (1680–1685?)." In: Schepers H., et al. (eds), *Sämtliche Schriften und Briefe*, Volume VI, Book IV. Berlin & Göttingen: Akademia Verlag, pp. 398–400.

Leibniz, G.W. (1685/1999) "Definitiones Notionum Metaphysicarum Atque Logicarum (1685)." In: Schepers H., et al. (eds), *Sämtliche Schriften und Briefe*, Volume VI, Book IV. Berlin & Göttingen: Akademia Verlag, pp. 624–30.

Leibniz, G.W. (1698/1998) "A Letter from M. Leibniz to the Editor, Containing an Explanation of the Difficulties Which M. Bayle Found with the New System of the Union of the Soul and Body." In: Francks, R., and Woolhouse, R.S. (trans), *G.W. Leibniz: Philosophical Texts*. Oxford: Oxford University Press, pp. 201–8.

Leibniz, G.W. (1703/1989) "Leibniz to de Volder, 20 June 1703 [excerpts]." In: Ariew, Roger, and Garber, Daniel (eds), *Philosophical Essays*. Indianapolis, IN and Cambridge: Hackett Publishing Company, 174–8.

Leibniz, G.W. (1704/1989) "Leibniz to de Volder, 1704 or 1705." In: Ariew, Roger, and Garber, Daniel (eds), *Philosophical Essays*. Indianapolis, IN and Cambridge: Hackett Publishing Company, pp. 181–4.

Leibniz, G.W. (1712/1989) "Notes for Leibniz to Des Bosses, 5 February 1712." In: Ariew, Roger, and Garber, Daniel (eds), *Philosophical Essays*. Indianapolis, IN and Cambridge: Hackett Publishing Company, pp. 199–200.

Leibniz, G.W. (1712/2007) "Leibniz to Des Bosses, 16 June 1712." In: Look, Brandon C., and Rutherford, Donald (eds), *The Leibniz-Des Bosses Correspondence*. New Haven, CT: Yale University Press, pp. 255–7.

Leibniz, G.W. (1714/1989a) "Monadology." In: Ariew, Roger, and Garber, Daniel (eds), *Philosophical Essays*. Indianapolis, IN and Cambridge: Hackett Publishing Company, pp. 213–25.

Leibniz, G.W. (1714/1989b) "Remarks on Berkeley's *Principles*." In: Ariew, Roger, and Garber, Daniel (eds), *Philosophical Essays*. Indianapolis, IN and Cambridge: Hackett Publishing Company, p. 307.

Leibniz, G.W. (1989) *Philosophical Essays*, Ariew, Roger, and Garber, Daniel (eds), Indianapolis, IN: Hackett Publishing Company.

Leibniz, G.W. (1998) *G.W. Leibniz: Philosophical Texts*, Francks, R. and Woolhouse, R.S. (trans.). Oxford: Oxford University Press.

Leibniz, G.W. (2007) *The Leibniz-Des Bosses Correspondence*, Look, Brandon C., and Rutherford, Donald (eds). New Haven, CT: Yale University Press.

Lewis, David (1973) *Counterfactuals*. Malden: Basil Blackwell.

Lewis, David (1986a) *On the Plurality of Worlds*. Oxford: Blackwell.

Lewis, David (1986b) *Philosophical Papers, Vol. 2*. New York: Oxford University Press.

Loar, Brian (2003) "Phenomenal intentionality as the basis of mental content." In: Hahn, M., and Ramberg, B. (eds), *Reflections and Replies: Essays on the Philosophy of Tyler Burge*. Cambridge, MA: MIT Press, pp. 229–58.

Locke, John (1690/1979) *An Essay Concerning Human Understanding*. Nidditch, Peter H. (ed.). Oxford: Clarendon Press.

Lockwood, Michael (1993) "The grain problem." In: Robinson, Howard (ed.), *Objections to Physicalism*. Oxford: Clarendon Press, pp. 271–91.

Lorentz, Hendrik (1914) *Das Relativitätsprinzip: Drei Vorlesungen gehalten in Teylers Stiftung zu Haarlem*, Keesom, W.H. (ed.). Leipzig and Berlin: B.G. Teubner.

Lorentz, Hendrik (1916) *The Theory of Electrons and Its Applications to the Phenomena of Light and Radiant Heat: A Course of Lectures Delivered in Columbia University, New York, in March and April 1906*. Leipzig: B.G. Teubner.

Lorentz, Hendrik (1937) "The relative motion of the Earth and the ether." In: Zeeman, P., and Fokker, A.D. (eds), *Collected Papers*. The Hague: Nijhoff, pp. 219–23.

Lowe, E.J. (1996) *Subjects of Experience*. Cambridge: Cambridge University Press.

McDowell, John (1994) *Mind and World*. Cambridge: Harvard University Press.

McGinn, Colin (1995) "Consciousness and space." In: Shear, Jonathan (ed.), *Explaining Consciousness – The 'Hard Problem'*. Cambridge, MA: MIT Press, pp. 97–108.

McGurk, Harry, and Macdonald, John (1976) "Hearing lips and seeing voices." *Nature* 264, December 23/30: 746–8.

Mach, Ernst (1919) *The Science of Mechanics: A Critical and Historical Account of its Development*, McCormack, T.J. (trans.). Chicago, IL: Open Court Publishing Company.

Mackie, David (1999) "Personal identity and dead people." *Philosophical Studies* 95, 3: 219–42.

McTaggart, John Ellis (1908) "The unreality of time." *Mind* 17, 68: 456–74.

Margenau, Henry (1961) *Open Vistas: Philosophical Perspectives of Modern Science*. New Haven, CT: Yale University Press.

Meehl, Paul E. (1991) "The compleat autocerebroscopist: a thought-experiment on Professor Feigl's mind–body identity thesis." In: *Paul E. Meehl: Selected Philosophical and Methodological Papers*. Minneapolis, MN: University of Minnesota Press, pp. 184–248.

Mermin, N. David (1968) *Space and Time in Special Relativity*. New York: McGraw Hill.

Mermin, N. David (1985) "Is the moon there when nobody looks? Reality and the quantum theory." *Physics Today*, April: 38–47.

Merricks, Trenton (1998) "There are no criteria of identity over time." *Noûs* 32: 106–24.

Merricks, Trenton (2001) *Objects and Persons*. Oxford: Oxford University Press.

Michelson, Albert A., and Morley, Edward W. (1887) "On the relative motion of the Earth and the luminiferous ether." *American Journal of Science* 34, 203: 333–45.

Mill, John Stuart (1865/1979) An Examination of Sir William Hamilton's Philosophy, and of the Principal Philosophical Questions Discussed in his Writings. Toronto: University of Toronto Press.

Miller, Izchak (1984) *Husserl, Perception, and Temporal Awareness*. Cambridge, MA: MIT Press.

Minkowski, Hermann (1909/1952) "Space and time." In: *The Principle of Relativity*. New York: Dover Publications, pp. 73–91.
Newton, Isaac (1686/1934) *Mathematical Principles of Natural Philosophy, Volume One: The Motion of Bodies*, Cajori, F. (trans.). Berkeley, CA: University of California Press.
O'Shaughnessy, Brian (2000) *Consciousness and the World*. Oxford: Clarendon Press.
Olson, Eric T. (1997) *The Human Animal: Personal Identity Without Psychology*. New York: Oxford University Press.
Parfit, Derek (1971) "Personal identity." *Philosophical Review* 80, 1: 3–27.
Parfit, Derek (1984) *Reasons and Persons*. Oxford: Clarendon Press.
Parsons, Charles (1964) "Infinity and Kant's conception of the 'possibility of experience.'" *Philosophical Review* 73, 2: 182–97.
Pauli, Wolfgang (1958) *Theory of Relativity*, Field, G. (trans.). New York: Dover Publications.
Peirce, Charles S. (1892) "The law of mind." *Monist* 2, 4: 533–59.
Peirce, Charles S. (2009) "Critical Review of Berkeley's Idealism." In: Wiener, Philip P. (ed.), *Charles S. Peirce: Selected Writings*. New York: Dover Publications, pp. 73–88.
Pelczar, Michael (2009) "Content internalism about indexical thought." *American Philosophical Quarterly* 45, 2: 95–104.
Pelczar, Michael (2014) "Physical time, phenomenal time, and the symmetry of Nature." In: Oaklander, L. Nathan (ed.), *Debates in the Metaphysics of Time*. London: Bloomsbury, pp. 131–48.
Pelczar, Michael J., Krieg, Noel R., and Chan, E.C.S. (1993) *Microbiology: Concepts and Applications*. 6th edn. New York: McGraw-Hill.
Perry, John (1976) "The importance of being identical." In: Rorty, A.O. (ed.), *The Identities of Persons*. Berkeley, CA: University of California Press, pp. 67–90.
Phillips, Ian (2010) "Perceiving temporal properties." *European Journal of Philosophy* 18, 2: 176–202.
Phillips, Ian (2011) "Indiscriminability and experience of change." *Philosophical Quarterly* 61, 245: 808–27.
Poincaré, Henri (1913) "The measure of time." In: Cattell, J.M. (ed.) *The Foundations of Science, Volume I of Science and Education: A Series of Volumes for the Promotion of Scientific Research and Educational Progress*. New York: The Science Press, pp. 223–34.
Popkin, Richard H. (2003) *The History of Scepticism: From Savanarola to Bayle*. Oxford: Oxford University Press.
Price, H.H. (1940) *Hume's Theory of the External World*. Oxford: Clarendon Press.
Price, Huw (1996) *Time's Arrow and Archimedes' Point: New Directions for the Physics of Time*. Oxford: Oxford University Press.
Putnam, Hilary (1973) "Meaning and reference." *Journal of Philosophy* 70, 19: 699–711.
Putnam, Hilary (1982) "Brains in a vat." In: *Reason, Truth, and History*. Cambridge: Cambridge University Press, pp. 1–21.
Ramsey, F.P. (1925) "Epilogue." In: *The Foundations of Mathematics and Other Logical Essays*. London: Routledge & Kegan Paul, pp. 287–92.
Rashbrook, Oliver (2013) "An appearance of succession requires a succession of appearances." *Philosophy and Phenomenological Research* 87: 584–610.
Robinson, Howard (1994) *Perception*. London: Routledge.

Rovelli, Carlo (2008) "Forget time." <http://arxiv.org/abs/0903.3832> [accessed June 12, 2014].

Rowe, William R. (1975) *The Cosmological Argument*. Princeton, NJ: Princeton University Press.

Russell, Bertrand (1910–1911) "Knowledge by acquaintance and knowledge by description." *Proceedings of the Aristotelian Society* 11: 108–28.

Russell, Bertrand (1914) "The world of physics and the world of sense." In: *Our Knowledge of the External World*. London: George Allen & Unwin, pp. 106–34.

Russell, Bertrand (1918) "The philosophy of logical atomism." *The Monist* 28: 495–527.

Savitt, Steven (2009) "The transient nows." In: Myrvold, Wayne C., and Christian, Joy (eds), *Quantum Reality, Relativistic Causality, and Closing the Epistemic Circle: Essays in Honour of Abner Shimony*. New York: Springer, 97–108.

Scerri, Eric (2009) *Selected Papers on the Periodic Table*. London: Imperial College Press.

Schaffer, Jonathan (2009) "On what grounds what." In: Chalmers, David J., Manley, David, and Wasserman, Ryan (eds), *Metametaphysics: New Essays on the Foundations of Ontology*. Oxford: Clarendon Press, pp. 347–83.

Schwitzgebel, Eric (2008) "The unreliability of naive introspection." *The Philosophical Review* 117, 2: 245–73.

Shoemaker, Sydney (1984) "Personal identity: a materialist's account." In: Shoemaker, Sydney, and Swinburne, Richard (eds), *Personal Identity*. Oxford: Basil Blackwell, pp. 67–132.

Sider, Theodore (2011) *Writing the Book of the World*. Oxford: Clarendon Press.

Skorupski, John (1994) "J.S. Mill: logic and metaphysics." In: Ten, C.L. (ed.), *The Nineteenth Century*. London: Routledge, pp. 98–121.

Smart, J.J.C. (1963) *Philosophy and Scientific Realism*. London: Routledge & Kegan Paul.

Soteriou, Matthew. (2007) "Content and the stream of consciousness." *Philosophical Perspectives* 21: 543–68.

Soteriou, Matthew (2010) "Perceiving events." *Philosophical Explorations: An International Journal for the Philosophy of Mind and Action* 13, 3: 223–41.

Stoljar, Daniel (2010) *Physicalism*. New York: Routledge.

Strawson, Galen (1994) *Mental Reality*. Cambridge, MA: MIT Press.

Strawson, Galen (1999) "The self and the SESMET." *Journal of Consciousness Studies* 6, 4: 99–135.

Strawson, Galen (2006) "Realistic monism: why physicalism entails panpsychism." In: Freeman, Anthony (ed.), *Consciousness and Its Place in Nature: Does Physicalism Entail Panpsychism?* Exeter: Imprint Academic, 3–29.

Strawson, Galen (2008) *Real Materialism: And Other Essays*. Oxford: Clarendon Press.

Swinburne, Richard (1984) "Personal identity: The dualist theory." In: Shoemaker, Sydney, and Swinburne, Richard (eds), *Personal Identity*. Oxford: Basil Blackwell, pp. 1–66.

Tetrode, Hugo (1922) "Über den Wirkungszusammenhang der Welt: Eine Erweiterung der klassischen Dynamik." *Zeitschrift für Physik* 10, 1: 317–28.

Thomson, William (Lord Kelvin) (1904) *Baltimore Lectures on Molecular Dynamics and the Wave Theory of Light: Founded on Mr. A.S. Hathaway's Stenographic Report of Twenty Lectures Delivered in Johns Hopkins University, Baltimore, in October 1884: Followed by Twelve Appendices on Allied Subjects*. Cambridge: Cambridge University Press.

Tooley, Michael (1997) *Time, Tense, and Causation*. Oxford: Oxford University Press.
Tye, Michael (2003) Consciousness and Persons. Cambridge, MA: MIT Press.
Valberg, J.J. (1992) *The Puzzle of Experience*. Oxford: Clarendon Press.
Vinten-Johansen, Peter, Brody, Howard, Paneth, Nigel, Rachman, Stephen, and Rip, Michael (2003) *Cholera, Chloroform, and the Science of Medicine: A Life of John Snow*. Oxford: Oxford University Press.
Wedeen, Van J., Rosene, Douglas L., et al. (2012) "The geometric structure of the brain fiber pathways." *Science* 335, 6076: 1628–34.
Wittgenstein, Ludwig (1958) *Philosophical Investigations*. 3rd edn. New York: Macmillan.
Wright, Crispin (1975) "On the coherence of vague predicates." *Synthese* 30, 3/4: 325–65.
Zuboff, Arnold (1977–78) "Moment universals and personal identity." *Proceedings of the Aristotelian Society* 78: 141–55.
Zuboff, Arnold (1981) "The story of a brain." In: Hofstadter, Douglas R., and Dennett, Daniel C. (eds), *The Mind's I*. New York: Basic Books, pp. 202–12.

Index

abominable conjunctions 164
absolute space 45–6, 216, 223
Adams, Robert M. 210
anti-realism 129–30
appearance, *see* experience "of" vs. "as of"
apple tree 181–2
argument from unusual origins 161–5
Aristotle 44, 119, 224
Armstrong, David 126, 128–9, 137, 139, 176
Arntzenius, Frank 221
astrolabe vs. microscope 175
atemporalism, *see* conscious experience: atemporalism
atomic experience 104–6
Ayer, A.J. 126, 139

backwards causation 221
bacon slicer 187
Barbour, Julian 6
bats 163
Bayne, Tim 11, 77–9, 182, 186
Bell, John S. 216
Berkeley, George 113–16, 122, 140
Berkovitz, Joseph 221
Big Bang 98
Block, Ned 17
Bohr, Niels 220
Boscovich, Roger Joseph 209
Brahe, Tycho 173
brains-in-a-vat 160–5
Brentano, Franz 178, 195
Broad, C.D. 91
Brown, Kevin 40, 44, 216, 221
Burge, Tyler 201

Callender, Craig 6, 8, 98, 221
Campbell, Keith 211
Campbell, Scott 185
Canberra Plan 212
Cargile, James 162, 193, 208, 216
Carnap, Rudolf 1, 140
Casati, Roberto 146
causal loops 221
Chalmers, David 11, 45, 77–9, 82, 112, 135, 140, 157, 162–5, 210–11
Chinese Brain 17
Chisholm, Roderick 192
Churchland, Paul 169, 175
Collins, John M. 164
concurrent roads 102
confounding qualia 13
conscious experience
  a well-founded phenomenon 59–62
  atemporalism 5, 29, 70, 110–11, 139–40
  atomic 104–6
  derivative location of 158–9
  diachronically complex 67, 82, 94–5, 99–101, 106
  diachronically simple 59, 69–71, 76, 80–2, 85, 94, 98–9
  in spacetime if in time 71
  interpretation of 117–20
  molecular 104
  temporalism 5, 79–82, 111
  unity of 74–9
consciousness, *see* conscious experience
cosmic microwave background 72
Craig, William Lane 215
Crane, Tim 200–2
cremation ashes 188
Crowther, Thomas 73
curved space 141

Dainton, Barry 79, 85, 87, 89–94, 98–101, 182, 186–8, 191
deceptive appearances 160, 165–7
Democritus 209
DeRose, Keith 164
Descartes, René 208
diachronic co-consciousness 94
diachronically complex experiences, *see* conscious experience: diachronically complex
diachronically simple experiences, *see* conscious experience: diachronically simple
direct realism 137
disunity problem 28, 75–9
doppelgangers 200
dreams, *see* deceptive appearances
Dretske, Fred 9, 164, 195
dualism 79–82, 114, 210–11
Dummett, Michael 130
duration, *see* physical time

Earman, John 149
Edelglass, William 192
Einstein, Albert 1, 44, 216

Einstein synchronization 38
embodiment 157–9
epiphenomenal ectoplasm 113
epistemic closure 164
EPR correlations 216–22
evolutionary psychology 6–7
experience, *see* conscious experience
experience "of" vs. "as of" 8–9
experience-holism 105–6
experiential voyages 147–51
explanatory posits 210
externalism
  indexical arguments for 203–6
  natural kinds arguments for 200–3
  properly defined 199–200
extra-bodies problem 145–6, 156–7

Farkas, Katalin 195
Feynman, Richard 210
Field, Hartry 195
Fine, Arthur 220
FitzGerald, George 216
Flamsteed, John 173
Fodor, Jerry 195
forest and kitchen 85–7
Foster, John 93, 208, 210
frame of reference, *see* inertial coordinates
functionalism 75
Futch, Michael J. 140

Galilean relativity 45
Galilei, Galileo 39, 41, 173, 216
Galison, Peter 38
Garfield, Jay 192
Gauss, Carl Friedrich 141, 147, 154
general relativity 224
Geroch, Robert 44
ghosts 113
God 122, 214
Gödel, Kurt 149
Gould, Rupert 37
Graff, Delia 61
grain problem 80
grounding 112
Gustafsson, Johan E. 180

Hacking, Ian 175
Hájek, Alan 125
hallucination, *see* deceptive appearances
Hartshorne, Charles 140, 210
Hawthorne, John 164
Healey, Richard 81
hidden patterns in experience 169
Hoerl, Christoph 62, 90
Horgan, Terence 195
hors d'oeuvres 213
hostage situation 182
human brain 24
Hume, David 84, 132–3
Humean supervenience 211–12
Hurley, Susan 11
Husserl, Edmund 88
Huygens, Christiaan 209

idealism
  Berkeleyan 113–14
  holistic 116–18
  Kantian 120–2
  Leibnizian 122–4
  mental substrate 116
  Millian 124–6, 140–6
  narrow 116
  principal objections to 160
  spacetime and 139–40
illusion, *see* deceptive appearances
imperceptible things 167–76
inertia, *see* principle of inertia
inertial coordinates 40–4
intentionality 31–2, 194–206
internalism
  about indexical beliefs 204–6
  basic argument for 196–8
  properly defined 199–200
intrinsic vs. relational properties 198–9
introspection
  limitations of 10–11
  reliability of 14–20, 27
  skepticism about 20–7
  under ideal circumstances 13
  Visual Brain and 17–19
Introspection Principle 11–13
inverted spectrum 100

Jackson, Frank 20, 112, 211
Julius Caesar 158
jumping streams of consciousness 181–2

Kant, Immanuel 27, 57, 115–16, 120–2, 140, 167–8
Kirk, Robert 211
kitchen and forest 85–7
Kreh, Lefty 207
Kriegel, Uriah 13, 195

Langton, Rae 117, 121, 168
laws of experience 96–101, 135–6
laws of physics
  CPT-reversal invariance and 98
  time-reversal invariance and 97–8
Lee, Geoffrey 4–5, 223
Leibniz, G.W. 30, 59, 74–5, 112, 115–17, 122–4, 129, 130, 136, 139–40, 178–80, 197, 207
Leibniz's mill 74–5
Lewis, David 3, 134, 211–12

LIFO vs. FIFO 34–5, 53
lightlike separation 47
Loar, Brian 194–5, 197
Locke, John 57–8, 209
Lockwood, Michael 80
Lorentz, Hendrik 215–16, 223–4
Lowe, E.J. 192

McDowell, John 9
McGinn, Colin 71
McGurk effect 106
Mach, Ernst 168
Mackie, David 183
McTaggart, J.M.E. 56, 62
McTaggart's Principle 62–6
manifest vs. scientific image 174–5
maps 33–4
Margenau, Henry 168
Maxwell's laws 40, 174
Meehl, Paul E. 75
memory of experience 184–5
Mendeleev, Dmitri 172
mental representation, *see* intentionality
merely apparent bodies 156–7
Mermin, N. David 44, 216
Merricks, Trenton 192, 210
Michelson, Albert A. 49
microscope vs. astrolabe 175
Mill, J.S. 2, 27, 30, 56–7, 116–17, 124–6, 127–9, 136–7, 140–6, 151–2, 159, 186, 208–9, 214
Miller, Izchak 58
Minkowski, Hermann 44
misperception 9
modal realism 134
molecular experience 104
monads 116–17, 122–4, 129, 136, 139–40, 197
movie analogy 63
Mr Nakamura 197
Müller-Lyer illusion 166

narrow content, *see* speaker-meaning vs. word-meaning
necessitarianism 134
neo-Lorentzian mechanics 224
Network Time Protocol, *see* Einstein synchronization
Newton, Isaac 27, 36, 39–44, 172–3, 224
Nike array 40–4
Nolan, Daniel 212

O'Shaughnessy, Brian 73, 185
Olson, Eric T. 183

parallel experience production 182
Parfit, Derek 184–5, 189
Parsons, Charles 168
part-sharing 101–2
Pauli, Wolfgang 44
Peirce, C.S. 70, 176–7
Pelczar, Michael J. 170
permanent possibilities of sensation 124–6, 128–9, 141
Perry, John 185
phenomenal character 103
phenomenal dispositions 32, 197
phenomenal duration
  independent of physical duration 94–5
  lower bound on 60–1
  no threat to atemporalism 68
  phenomenological temporalism and 5–7
  relativistic duration and 5–7, 94–5
  ubiquity of 62–6
phenomenal gunk 59
phenomenal integration 86–7, 101, 106
phenomenal overlap 92–3, 101, 106
phenomenal simultaneity
  dualism and 81–2
  physical simultaneity and 70–9
  physicalism and 79–81
  relativistic simultaneity and 6, 69–83
phenomenal sorites 106–10
phenomenal succession
  ubiquity of 62–6
  vs. phenomenal passage 62
  vs. succession of experiences 56–68
phenomenalism
  direct realism and 136–7
  in Mill 124–6, 140–6
  intentionality and 194–8
  mind-dependence and 128–30
  radical phenomenalism 1–2, 30, 139–59
  superior to other idealist theories 127–38
  the mystery of consciousness and 130–6
phenomenological blindspots 13
phenomenological hypertext 214
phenomenological limit 154
phenomenological potential 114, 120–6, 140–1
phenomenological regress 64–6
Phillips, Ian 58, 61, 88, 185
photographs 117–18
physical time
  as the possibility of self-difference 139–40
  defined 7
  duration in 5–6, 52–3
  simultaneity in 6, 36–8, 49–52, 215–24
physicalism 79–82, 114, 210–13
pickle jar 85–7
plague 168–9
Poincaré, Henri 3–7, 54–5, 111, 215–16, 221
pointillism 5, 79–82
Popkin, Richard H. 209
pre-established harmony 116–17, 122
Price, H.H. 143
Price, Huw 98, 221

principle of inertia 39, 223–4
principle of sufficient reason 131–4
problem of repeated contents 89–94, 101
protophenomenalism 210–11
Putnam, Hilary 196–7, 201–5

qualia 8–10

radical phenomenalism
  detailed development of 139–59
  global 151
  local 154
  metaphysical objections to 160–77
  motivation for 3–7, 28–9, 140, 208–14
  phenomenological objections to 2, 110–11
Ramsey, Frank 126, 213–14
random (vs. uncaused) 135
Rashbrook, Oliver 58, 73
relational vs. intrinsic properties 198–9
relativity, *see* special relativity
relativity of duration 52–3
relativity of simultaneity 49–52
risks 129
Robinson, Howard 113, 124, 125, 208
Rovelli, Carlo 8
Rowe, William R. 133
Russell, Bertrand 5, 69–70, 126, 192–3, 195
Russell's Thesis 69–73

Savitt, Steven 6
Scerri, Eric 172
Schaffer, Jonathan 112
Schwitzgebel, Eric 20–6
scientific realism, *see* imperceptible things
scientific vs. manifest image 174–5
Second Law of Thermodynamics 98
self
  anti-realism about 180
  asymmetric dependence theory 181, 184–6
  Buddhist views on 192
  hypothetical-stream theory 186–8
  particularism about 189–92
  reductive theories of 181–8
  substrate theory 183–4
Sellars, Wilfrid 174–5
Semmelweis, Ignaz 170
sensation-conditionals 125, 141–6
serialism 28, 70, 73–9
Shearing, George 17–18
Shoemaker, Sydney 185
Sider, Theodore 212
simultaneity, *see* physical time
skepticism 164, 208–9
Skorupski, John 125, 209
Smart, J.J.C. 9, 128, 130–1, 143–4
Snow, John 170–1
soda machine 183
Soteriou, Matthew 58, 64, 73
spacelike separation 47
spacetime
  distance in 54, 210
  topology of 148–50
  violation of triangle inequality 210
spacetime explorer 146–7
speaker-meaning vs. word-meaning 203
special relativity 44–54
spherical interstellar object 133–4
Stoljar, Daniel 112
Strawson, Galen 73, 180, 195, 210
stream of consciousness
  atemporal theory 101–6
  divided appearances in 89–90
  inner-awareness theory 91–2
  memory theory 90
  neutrally characterized 84
  paradox of 87–9
  temporal overlap theory 92–4
Strobe Case 16
supervenience 112–13

telegraph 37–8
temporalism, *see* conscious experience: temporalism
Tetrode, Hugo 33, 221
Texas to Nevada 119
Theorema Egregium 141
things-in-themselves 120–2, 130, 136
Thomson, William (Lord Kelvin) 209
Tienson, John 195
time travel 146
timelike separation 47
time-slices 44
Tooley, Michael 223
Tree of Science 210
Turing machine 123; *see also* monads
Turner landscape 97
Twin Earth 200, 222
two brains 182
Tye, Michael 85

uncaused (vs. random) 135
unity of consciousness 73–9
Unworld 98–100
Uttaklak and the Eelgoo 144–5

Valberg, J.J. 56, 125
van Fraassen, Bas 175
van Inwagen, Peter 131
Varzi, Achille C. 146
Vatmaster 163
Venus 18–19
veridical perception 9
verificationism 216
Victor 18–19

Vinten-Johansen, Peter 171
Visual Brain 18–19, 60

"water"-dispositions 202
waterfall illusion 8, 63
Weierstrass function 64–5
Wheeler-Feynman absorber theory 221
wide content, *see* word-meaning vs. speaker-meaning
William of Ockham 207
Winkler, Kenneth 143
Winter Solstice 67
word-meaning vs. speaker-meaning 203
Wright, Crispin 61

Zuboff, Arnold 5, 95

The manufacturer's authorised representative in the EU for product safety is Oxford University Press España S.A. of El Parque Empresarial San Fernando de Henares, Avenida de Castilla, 2 - 28830 Madrid (www.oup.es/en or product.safety@oup.com). OUP España S.A. also acts as importer into Spain of products made by the manufacturer.
Printed and bound by CPI Group (UK) Ltd, Croydon, CR0 4YY
14/09/2026
02216308-0009